SOCIAL DIMENSIONS OF WATER SUPPLY AND SANITATION IN RURAL AREAS

About the Author

Professor Sachchidananda, currently Director of the Sulabh Institute of Development Studies, has been a pioneer in the field of Anthropological studies in Bihar. He was educated in the Patna University and the School of Oriental and African Studies, University of London. Founder of the Post-graduate Department of Anthropology at Ranchi, he was later Director of the Bihar Tribal Research Institute, Ranchi and the A. N. Sinha Institute of Social Studies at Patna. He was also the Vice-Chancellor of Ranchi University. His wide ranging research interests cover culture change in tribal areas, shifting cultivation, social dimensions of agricultural development, Harijan educated elite, women's rights, educational problems of the disadvantaged sections of the population, voluntary action in rural areas and socio-cultural dimensions of development. He has contributed a large number of papers to research journals in India and abroad and published twenty books in Sociology and Anthropology. He has also edited the four volume *Profile of Indian Tribes* (1995). A past President of the Anthropology Section of Indian Science Congress, he is also the recepient of R.P. Chanda and S. C. Roy Medals for distinguished work in Anthropology from the Asiatic Society of Bengal.

SOCIAL DIMENSIONS OF WATER SUPPLY AND SANITATION IN RURAL AREAS

A Case Study of Bihar

PROFESSOR. SACHCHIDANANDA
M.A. (Patna) M.A. (London), D. Litt.
Director, Sulabh Institute of Development Studies, Patna
Former Director, A.N. Sinha Institute of Social Studies, Patna
and
Former Vice-Chancellor, Ranchi University

Assisted by
DR. R. K. PANDEY
Research Officer,
Sulabh Institute of Development Studies

Published for

SULABH INSTITUTE OF DEVELOPMENT STUDIES

by

CONCEPT PUBLISHING COMPANY, NEW DELHI-110059

ISBN 81-7022-723-2

First Published 1999

Published and Printed by
Ashok Kumar Mittal
Concept Publishing Company
A/15-16, Commercial Block, Mohan Garden
New Delhi 110059 (India)

Phones: 5648039, 5649024
Fax: 091-(11)-5648053

PREFACE

Low access to the basic health care facility is a common phenomenon in most developing countries. In 1978, it was decided to launch a movement throughout the world known as 'Health for All' by the year 2000 A.D. India is a signatory to that declaration and since then efforts have been made to extend health facility to all so that it reaches the vulnerable sections in the society. Safe drinking water supply and basic sanitation are vital human needs for health and efficiency. Disease and death, particularly in children every year and the drudgery of women are directly attributable to lack of these essentials.

The National Water Policy in 1987 gave the highest priority to drinking water supply. However, sanitation, both in urban and rural areas, continues to remain neglected. Both rural water supply and rural sanitation became important constituents of the state sector Minimum Needs Programme. Highest priority was to given to provide safe drinking water to problem villages. It was to be ensured that the scheduled caste and scheduled tribes population and other weaker sections community are fully covered. All existing dry latrines were to be converted into low cost sanitary latrines. In order to reduce the cost, the programme was to be implemented in a decentralised manner with the involvement of the people.

In view of the clear cut policy directives and the provision of funds for this purpose both by the Central and State Government, it was expected that considerable progress in this direction would have been made in Bihar during the last ten years. This study was undertaken to highlight the social dimensions of rural water supply and sanitation in the context of the existing situation in this state. An effort was made to study the role of such social variables as region, sex, education, caste, religion, income and exposure to media. The constraints in achieving results emanating from behaviour, habits, attitudes and belief in the provision of safe drinking water and sanitation facilities in rural areas were also explored. The views of women and the youth in different cultural zones in the state on these two issues were recorded. The contribution of voluntary agencies in these two fields were also examined. Towards the end of the study some suggestions have been made for hygiene education and social mobilization of the people for the adoption of low cost toilets.

During this study data was collected from 130 villages spread over 13 districts, one from each division in the state. Inter-cultural comparisons have been made as it has been realised that the entire issue has to be viewed from a socio-cultural point of view rather than only from the point of technical feasibility. The gender perspective has been given salience in the study.

I am grateful to Dr. Rajkumar Pandey, Research Officer, Sulabh Institute of Development Studies, Patna for helping me right from the planning stage. The three Research Assistants involved in the study deserve hearty thanks for doing the field work sincerely and diligently.

The Sulabh International Social Service Organisation funded this project. I am thankful to the then Chairman of the Bihar Branch Sri S.N. Jha for taking interest in the study and for making some very useful books, reports and papers from his personal library available to us.

I am deeply obliged to Shri Ashok Kumar Mittal of the Concept Publishing Company for seeing the study through the press in a short time.

Sachchidananda
Director
Sulabh Institute of Development
Studies, Patna.

CONTENTS

LIST OF TABLES

1

SAFE DRINKING WATER, SANITATION AND SUSTAINABLE DEVELOPMENT

Safe Drinking Water

Water is basic need of life. Safe drinking water could make a tremendous impact on the quality of life in our villages, because availability of safe drinking water has direct impact on the working conditions and health of the people and their productive capacity. It has been noticed that the problem of supply of potable water to a very large population is very complex and challenging.

It is impractical to assign precise monetory values to the debilitating effect of water related diseases. The effects encompass the decrease in labour productivity, the increase in public cost of therapy and the offsetting costs of public health preventive measures. Rough measures of these can be made by estimating the degree to which water improvement reduce the disease incidence from the maximum which would prevail with scarce and heavily polluted water supplies.

Although in general terms better and more water leads to better health, the particular evidence supporting this conclusion is complex and not readily summarized. It is helpful to consider water related disease in four categories. Water borne diseases in the strict sense are infections carried by polluted water, give rise to epidemics and are removed by providing pure water. The well-known infections, such as typhoid and cholera, may be augmented by various other infections where pollution reaches very high levels. Another large group of diseases, which we have called water-washed, are reduced when ample water is available for hygiene even if that water is not completely pure. Many of these are superficial infections of the skin and eyes. The other large group in this second category is the diarrhoea disorders. A group of tropical parasitic infections, which we may call water based, depend on quatic intermediate hosts for development. The larval parasites may either bore in through the skin on with schistosomes or be ingested. The fourth category of infections related to water includes those with insect vactors that either breed in water or bite near surface water. All four categories should be considered in any alteration of water resources—it is not unusual for improvements in some aspects of supply to create additional health problems.

Sanitation

The term sanitation has evidently a connotation not limited to latrines, embraces personal hygiene, community well being, garbage management, low cost housing, water supply, drainage system refuge collection and environmental health in general. The latrine which is so crucial a component of sanitation that we legitimately use the terms synonymously and tend to identify one with the other. The close connection between rural latrines and rural sanitation underscores the need to pursue the low cost latrine programme in rural areas with dedication, if we are serious about improved sanitation.

Sustainable Development

The time honoured debate about what is meant by development has perhaps best been summed by the

Administrator of United Nations Development Programme (UNDP) as follows:

> "Sustainable human development is development that not only generates economic growth but distributes its benefits equitably; that regenerates the environment rather than destroying it, that empowers people rather than marginalising them. It gives priority to the poor, enlarging their choices and opportunities, and provides for their participation in decisions affecting them, it is development that is pro-poor, pro-nature, pro-jobs, pro-democracy, pro-women and pro-children."

Sustainable development is a multi-dimensional concept with three interacting angles—ecology, economics and ethics. The necessary conditions for achieving sustainable development are ecological security, economic efficiency and social equity. Sustainable development does not end with the sustainability of just the environment and resource system but requires the sustainability also of economic and social system. One of the most important sectors for development has been the provision of potable water supplies, Sanitation and hygiene education to urban, peri-urban and rural communities. Improvements in those areas are essential to promote health for the general population and for child survival in particular and are fundamental for the development of many industries and business and overall urban expansion. Major water supply and sanitation facilities especially those serving urban areas and economic centres form part of the national infrastructure like roads and electric power grids. In rural areas, improving water supplies, providing adequate sanitation and encouraging safe hygiene practices are intrinsic to community development.

Water supply and sanitation are fundamental building blocks in the development process, influencing economic development, employment, agriculture, housing, health and numerous other sectors. These benefits are spread broadly across, societal concerns. However, support and financing for water supply and sanitation projects are usually most effectively justified on health grounds. As Dr. Halfdan Mahler, former Director General of the World Health Organisation (WHO) has said, "The number of water taps per 1000 persons is a better indicator of health than the number of hospital beds."

Health benefits result from improved quality and increased quantities of water, adequate sanitation facilities, and changes in hygiene behaviour. For a number of waterborne diseases (e.g. dracunculiasis and cholera) transmission occurs as a direct result of drinking contaminated water. In other cases, the transmission cycle is through contact (e.g. Schistosomisis) water also provides a breeding site for disease carrying vactors (e.g. malaria-bearing mosquitoes). Inadequate water for washing hands and cooking utensils and poor sanitation practices often result in diarrhoeal diseases. Disease and poor health exact a heavy burden on the economy, the community, the family and the individual. Improvements in water supply and sanitation can reduce the incidence of many of these diseases or eliminate them entirely.

Study after study indicates that children benefit greatly from improvements in water supply and sanitation. Improved water supplies and sanitation practices are instrumental in reducing infant mortality, preventing diarrhoea, and improving child nutrition and overall health, and this fact can and should be used as a powerful lever in garnering support for water supply and sanitation improvements.

In addition, unlike many other types of health interventions, water supply and sanitation improvement result in a whole range of secondary non-health benefits. They provide economic benefits for the population as a whole and for women in particular by reducing the time spent in getting water. Unfortunately, it can be difficult to demonstrate the many benefits of improved water and sanitation conclusively through conventional cost-effectiveness techniques. Water supply and sanitation facilities are necessary but not always sufficient by themselves to achieve substantial disease reduction. Rather than showing an immediate and direct impact on disease reduction, water supply and sanitation facilities appear to have a long run "multiplier" effect that reduces mortality as a result of environmental improvement. One of the few sets of data available corroborate this effect. It indicates that in urban France in the 19th century, sharp drop in mortality patterns coincided with the introduction of piped water and sewer systems. Evidence also suggests that negative health effects multiply upon the decline of adequate water and sanitation. For example, the return of cholera to Latin America in 1991 may be seen as an indication that water and sanitation conditions are worsening as increased population and urbanisation stretch the resources of already poor countries to the breaking point.

Diarrhoeal diseases that result from contaminated water kill approximately 2 million children annually. Diarrhoea which also affects childhood nutrition. The primary cause of growth failure in developing country children is poor nutrition, resulting in part from frequent bouts of diarrhoea which inhibit the ability of the body to absorb food. Thus children already at risk of dying because of the effects of diarrhoea are also at risk nutritionally. Diarrhoea, along with shortage of food, keeps growth rates below standard for most poor children despite medical and nutritional interventions.

Water Related Health Problem

Of the thirty-seven major diseases in developing countries twenty-one are water and sanitation related. For ten of these twenty-one, water and sanitation are considered primary interventions for prevention; diarrhoea, typhoid, schistosomiasis amebiosis, hookworm, hepatitis, ascariasis, giardiasis, trichriasis and drancunculiasis. These ten diseases cause a tremendous amount of illness every year. One source, using 1986 figures, states that the ten diseases are responsible for over 28 billion disease episodes a year in developing countries. According to the 1993 World Bank's World development report, in 1990 mortality from the ten diseases was approximately 3,000,000, but the big killer among them was diarrhoeal disease with a total of 2,866,000 deaths.

The Importance of Sanitation

Sanitation should be accorded the same priority as water supply. The connection between water supply and sanitation is implied by the name of the sector, but the sanitation half of the equation has too often been ignored. Water supply and sanitation development are inextricably related from a health standpoint and, although improvements in either have the potential for positive health effects, their combined impact is greater than the sum of the two parts,

Sanitation refers principally to the safe handling, treatment and disposal of excreta, although the handling, treatment disposal, and industrial waste are related and equally serious functions. Hygiene practices, such as the safe storage and handling of drinking water, also are a part of sanitation.

The plan of action for the water supply and sanitation decade called for accelerated programmes in both areas. However water supply received far more emphasis than sanitation and population growth, especially in peri-urban areas. This made the imbalance even more pronounced. As a result, today there is a great need for sanitation which will be met only if sanitation is accorded much greater priority than it was during the decade. Many urban areas have sewage systems that take waste water away from dwellings, but the waste water is not treated. It is simply dumped in raw form in some nearby water course, often creating a health hazard for another community.

The sanitation problem is made more acute by the explosion of urban growth. Between 50 and 80 per cent of this growth is occurring in the informal sector in outlying squatter areas. People settle in these peri-urban areas before any infrastructure is in place because squatters often settle on the least desirable land steep hillsides, low-lying flood plains contaminated sites, etc. Neither municipalities nor individual landowners find it cost effective to provide infrastructure of any kind (roads, electricity, water or sanitation). Once poor people begin to settle on these sites because these are the only places they can afford, it becomes even more difficult and more expensive to install water mains or sanitation system. While the people living in these settlements eventually find ways to obtain water roads, electricity and schools, their demand for sanitation is likely to rank well below these other priorities. Yet the sheer volume of waste and waste-water produced in these densely settled areas is staggering and poses serious health and environmental risks.

Hygiene education is generally included in water supply projects, but improving sanitation facilities still receives far less emphasis in project development. To some extent, poor sanitation practices can be compensated for by increased quantities of available water for cleaning and hygiene, nevertheless, insanitary waste disposal is one of the leading causes of water contamination and of many water borne and water related diseases. Sanitation can be addressed more effectively by facilities improvements than by increases in the water supply alone.

The Water and Sanitation Health (WASH Project 1990) review of studies on the health effects of water and sanitation found that safe excreta disposal is the most effective intervention for reducing the incidence and severity of the six diseases studied. Using data from that review and another similar survey, the World Bank has estimated that two million diarrhoea induced deaths among children under five could be prevented annually through the provision of water and sanitation. The importance of sanitation to health has been confirmed by other studies as well. This study using demographic and health survey data from Guatemala was designed to test the hypothesis about the relative benefits of water and sanitation, rural/urban differences in such benefits, and the significance of community level compared with family level sanitation efforts. The study found that children who live in urban communities with poor sanitation were twice as likely to suffer from stunted growth from bouts of chronic diarrhoea as those who live in communities with high levels of sanitation. Even when they had no individual access to a toilet, children in communities with high levels of sanitation (where at least 75 per cent of the households had toilets) were less likely to be stunted than those living in communities with low levels less than 75 per cent). How everyone else in the community disposes of faeces is a significant factor in an individual's chances of being infected with disease.

The environmental problems resulting from the lack of sanitation, particularly in urban areas, are enormous. In peri-urban settlements, most facal matter ends up on the ground and then either washes down to the nearest river or gets diluted by rain. Thus these settlements, producing tonnes of facal matter, become the largest one-point source of pollution in the city, polluting the river, the ground, and the water table.

Sanitation is often assumed to be a prohibitively expensive intervention beyond the reach of most communities and impossible to achieve for all. But there is growing recognition that the costs of inaction are also high, that a range of technology choices is available, and that success depends at least as much on behavioural change and land ownership and politics as on specific technologies.

Impact on Health and Environment

Studies reveal that over 50 infections can be transmitted from a diseased person to a healthy one by various direct or indirect routes from human excreta. A large number of deaths in India are from diarrhoeal diseases. The impact of safe disposal of human excreta on health has been significant in reducing morbidity and mortality from gastro-enteric disease and helminthic infections. Out of 950 million people, 700 million either defecate in the open or use insanitary bucket or dry privies exposing the entire community to health hazards and foul environment. The number of bucket and dry privies in India is about 76.4 lakhs of which nearly 54 lakhs are in urban India.

Relationship between Water Supply and Sanitation

About 1.3 billion people in the developing world lack access to clean and plentiful water, and nearly 2 billion people lack an adequate system for disposing of their feces. Feces deposited near homes, contaminated drinking water (sometimes caused by poorly designed or maintained sewerage systems) fish from polluted rivers and coastal waters and agricultural produce fertilised with human waste are all health hazards. Water quantity is as important as water quality. Washing hands after defecation and before preparing food is of particular importance in reducing disease transmission, but without abundant water in or near the home, hygiene becomes difficult or impossible. The lack of water supply and sanitation is the primary reason why diseases transmitted via feces are so common in developing countries. The most important of these diseases, diarrhoea and intestinal worm infections, account for an annual burden of 117 million Disability Adjusted Life Years (DALYs), or 10 per cent of the total burden of the disease in developing countries. In addition, an inadequate water supply increases the risk of schistosomiasis, skin and eye infections and guinea worm disease.

Unlike earlier times, today sanitation is viewed as a comprehensive concept which includes liquid and solid waste disposal, food hygiene, personal, domestic as well as environmental hygiene. It is for the first time that the National Drinking Water Mission has prepared a massive information, education and communication strategy for creating awareness among the rural people in water and sanitation sector. It is encouraging to note that 65 districts in 12 states of the country have been taken up the strategy for intensive awareness campaign to promote behavioural changes in respect of sanitation practice and sanitary aspects of water supply.

Before water is consumed, it passes through several stages of transportation, storage and handling. Avoidable contamination occurs in this process.

Utensils are often not covered, hands are dipped too frequently and storage practices are improper. Misconception about the relationship between water and health are galore. Most people link bad drinking water erroneously with fever, cough, cold, malaria. etc. Less than 20 per cent of people surveyed by the cited study were aware of link between drinking unsafe water and diarrhoeal diseases which kill over 15 lakh children under five, every year.

The relative inability of the mission, among other factors is amply illustrated by the following summary of the "Report of the expert committee on Rural Water Supply Programme" appointed by the Govt. of India in April, 1994. The Committee was constituted to evaluate the Rural Water Supply Programme with special reference to the Mini Mission and sub-missions in 1995. The Committee visited five states and held elaborate discussions with senior officials and people's representatives as well as non-government organisations and beneficiary public. Several reports were perused and detailed information was collected through a questionnaire from the states. "The Mission mode" though strategically well-conceived at the apex level revealed several tactical shortcomings and failures at the implementation level. The strategy followed was adhoc and has not been institutionalised. The programmes could not achieve the integration of services and resources from different sectoral agencies to ensure sustainability. Several committees constituted to achieve integrated functioning failed and attempts at networking did not strike root. Also the mission mode of functioning has not provided the desired level of flexibility to enable quick decision making. Right from the stage of sanction of the scheme, important software aspects like training, awareness building, community participation, micro level ecological planning were not accorded due priority.

The scientific source finding method alone appears to have succeeded and all other aspects of sustainability relating to water conservation and water recharging have been neglected. The traditional technologies and systems have not been well integrated into rural water supply systems though they have been tried out sporadically.

Water is life and sanitation is a way of life. Both are critical for improvements in the quality of life and health, for 80 per cent of the diseases stem from consumption of unsafe water and poor sanitary condition in the country. Towards supplementing and capping the sanitation arrangements made through the 20 point programme and the MNP, a Central Rural Sanitation Programme (CRSP) was introduced in 1986 by the Government of India.

During the Seventh Plan and two Annual Plans, a sum of Rs. 320 crore was spent under the State and Central Sector Programmes. The percentage of rural population covered by sanitary facilities reached 2.8 per cent at the end of 1991-92 from about 0.8 per cent at the commencement of the Seventh Plan. However, the results of the 44th round of NSS (1988-89) showed that over 10 per cent of the rural population had access to sanitary facilities. This indicates that additionally 7 per cent of the population secured access to sanitation through private initiatives. The survey also observed that the percentage of rural population with sanitation facilities was much larger in states like Kerala.

Water and Sanitation In Development Countries

The essential paradox of community water supply in developing country like India is that in one sense, everyone has water supply while in another sense, most people do not have safe drinking water. All countries must have some kind of water source. It may be dirty, it may be inadequate in volume and it may be several hours walk away but neverthless, some water must be available. However, if we apply any reasonable criterion of adequecy in terms of the quantity, quality and availability of water then most people in developing countries do not have an adequate supply.

The World Health Organisation's figures for 1980 show that among the urban population of the developing countries, only about 55 per cent have house connections and an additional 20 per cent have access to public taps. About a half of the rural population have intermittent water supply, less than a quarter have access to safe water, and hardly any of these have house connections. Overall, 62 per cent of people in developing countries lack an adequate water supply. The people without adequate water supplies number at least 1200 million.

Excreta Disposal in Developing Countries

The World Health Organisation's figures for 1980 showed that only 53 per cent of the combined urban population of the developing countries had adequate facilities for excreta disposal. Only a minority of these were served by piped sewerage systems. In the rural areas only 13 per cent had adequate excreta disposal facilities. These figures are based on very modest definitions of adequacy. Many latrines do not meet minimal public health requirements are not accessible to children or are liable to pollute nearby wells.

It is estimated that over 171 million people are without a safe source of drinking water. According to the UNDP (1993) Human Development Report population with access to water supply in India during the year 1990 was 75 per cent. Some of the other developing countries like Brazil (96%), Turkey (92%), the Philippines (81%) and Bangladesh (78%) are much better than India in regard to access to water supply. The lack of safe drinking water creates an unfavourable environment for healthy living leading to all kinds of water borne disease again, it is also found that about 640 million people have no access to sanitation. The above mentioned reports also maintain that only 13 per cent people in India in 1990 had access to sanitation compared to 99 per cent in the case of the Korean Republic, 94 per cent in Malaysia, 96 per cent in China and so on.

Thailand happens to be the only country that was able to meet the objectives of providing safe drinking water to its rural population by the end of the United Nation, International Drinking Water Supply and Sanitation decade (1981-90). It was based on a common practice which has 2000 years old tradition in Thailand. Between 1983 and 1986 the Thai government conducted its Jar programme under which nine million jars were constructed and distributed amongst its 18 million rural inhabitants. The technique is old, the materials used for the Jars were modern.

Water Supply and Sanitation in India with Special Reference to Bihar

Safe drinking water supply and basic sanitation are vital human needs for health and efficiency. Disease and death, particularly of children every year and drudgery of women are directly attributed to lack of these essential. The requirement of drinking water is at present only 7 to 10 per cent of water available on the surface and 4 to 5 per cent of the ground water potential. The National Water Policy announced in 1987 gives the highest priority to drinking water supply. Sanitation, both in urban and rural areas continues to remain neglected. Funds availability has been a constraint in the urban areas while schemes for rural sanitation were not able to evoke response.

Rural water supply was an important constituent of the State sector MNP during the Seventh Plan. In 1986, the National Drinking Water Mission (NDWM) was launched in order to provide scientific and cost effective content to the centrally sponsored Acclerated Rural Water Supply Programme (ARWSP).

Rural Sanitation Programme was also added to the state sector for Minimum Needs Programme (MNP) from 1987-88. In November, 1986, a new Centrally Sponsored Rural Sanitation Programme (CSRSP) was launched.

Based on the perceptions gained from an analysis of the water supply and sanitation sector in the past and the overall approach and strategies adopted in New Delhi Declaration, the following objectives were visualised for the Eighth Five Year Plan.

(i) Highest priority has to be given to ensure that the remaining about 3,000 'No source' hard-core problem villages in some states are provided with sustainable and stipulated supply of drinking water by March 1993.

(ii) Equally important would be to ensure that all the partially covered villages having a supply level of less than 40 litres per capita per day numbering about 1.5 lakh including hamlets are fully covered with safe drinking water facilities by the end of the Eighth plan on sustainable basis.

(iii) Ensure that SC/ST population and other poor/ weaker sections are covered fully on a priority basis.

(iv) The stipulated norms of supply would be 40 litre per capita per day of safe drinking water within a walking distance of 1.6 kms. or elevation difference of 100 metres in hilly areas.

(v) Achieve coverage of about 5 per cent of rural population with their full involvement with sanitation facilities by the end of Eighth Plan. Information, Education and Communication (IEC) should be an integral part of rural sanitation programme. In this context, it is desirable to adopt the concept of 'total environmental sanitation' and provide guidelines to the rural population in regard to proper environmental sanitation practices including disposal of refuge, garbage and waste water through the mechanism of local village leaders and community organisations and construct biogas plants adjacent to the sanitary complexes.

(vi) Convert all existing dry latrines into low cost sanitation latrines. In order to reduce costs, it is desirable that the programmes of drinking water supply and sanitation, both in rural and urban areas, are implemented in a more decentrralised manner. With the involvement of the people and local institutions at all stages such as planning, project formulation, execution, operation and maintenance, monitoring and evaluation.

Thrust and Strategies

With regard to operation and maintenance the following aspects need to be emphasised:

(1) Water has to be managed as a commodity in exactly the same as any other resource.
(2) Supply of water to consumers should normally be based on the principle of effective demand which should broadly correspond to the standard of service, that the users are willing to maintain, operate and finance.
(3) In rural areas, Water tariff may not be feasible in all places. However, wherever house service connections are given, it is suggested that appropriate water tariff is levied and realised

where by operation and maintenance becomes self-sustaining to the extent possible.

(4) Appropriate links should be forged between water supply and environment sanitation in the planning of new programmes.

The concept of 'total sanitation' covering primary health care, water availability women's welfare immunization and provision of sanitation facilities, all linked to cleanliness as a basic human need will be emphasised. Every effort will be made to adopt a low-cost approach, employing technical and scientific knowledge and the experience already gained by several non-governmental organisation in this regard.

It is desirable that the district planning machinery is used to provide the necessary technical support to the local bodies for creation and maintenance of installation of available water for various uses. It is necessary to take timely steps to see that the sources are recharged and rejuvenated. It will at the same time, be necessary to ensure appropriate linkage between soil-conservation and land/water management and drinking water supply schemes in order to provide for sustained supply.

Most states face resource problems and, therefore, tend to neglect maintenance. It is necessary to evolve an effective mechanism for ensuring proper operation and maintenance of existing assets 'Village-water Committee' should be actively involved in the maintenance of drinking water supply schemes and a system of maintenance of drinking water supply schemes and a system of beneficiary participation introduced. Participation of village women and NGOs/ voluntary organisations should also be encouraged. The mechanism and the funds available under TRYSEM should be used to impart training, so that trained manpower can be locally mobilised for the maintenance of the assets.

After a careful consideration of the situation as regards rural sanitation, it was suggested that the Rural sanitation programme be restructured providing for the following elements.

(a) All activities under this programme would be undertaken through local body/village panchayat and with beneficiary participation.
(b) Wherever feasible, NGOs would be involved in the implementation of the programme.
(c) Women would be actively associated with the implementation of the programme.
(d) Some contribution would be solicited from the beneficiaries, at least in the form of physical labour, in order to engender among them the realisation that the assets created belong to the local community.

Rural water supply and rural sanitation facilities are essential ingredients in the total programme for rural development. There are many other allied elements which go into the process of rural development. These include infrastructure like land management, soil conservation, afforestation etc. and social aspects like primary health care, removal of illiteracy, women's welfare, child nutrition, immunisation etc.

(A) STATUS OF DRINKING WATER AND SANITATION IN RURAL INDIA

Water, an absolute necessity of life, can also prove to be a carrier of suffering and death if not properly treated and supplied in adequate quantity. At the beginning of the present Water Decade, 31 per cent of the rural population was provided with safe drinking water supply facilities. This does not mean that 69 per cent of the rural population did not have any water supply. In fact, it means that 69 per cent of rural population were without any safe drinking water supply available within reasonable distance and had to depend on water sources, such as rivers, wells and ponds most of which go dry during summer or are highly polluted and a risk to health. Most of these water sources are family owned but highly susceptible to pollution and thereby are potential health hazards.

On the other hand, the drinking water sources created through government are mostly tapping sustainable sources available throughout the year and designed to provide protection against pollution. The rural sanitation programme was taken up in January 1986 under the National Rural Employment Programme (NREP), the Rural Landless Employment Guarantee Programme (RLEGP), Indira Awas Yojana and a number of other schemes. In the States in order to supplement the efforts of the State Governments, the Centrally Sponsored Rural Sanitation Programme (CRSP) was launched in October 1986. The Rural Sanitation Programme was included under the Minimum Needs Programme (MNP) in 1987-88. The Rural Water Supply and Sanitation Programmes are two important elements in the Government of India's 20-point programme for socio-economic development. In the later half of the National Seventh Five Year Plan (1985-90) UNICEF also came forward to support the government's efforts in the provision of rural sanitation. In eight states of India, while the urban water supply and sanitation programmes are implemented under the state plans a centrally Sponsored Low Cost Sanitation Programme was launched in 1990-91.

Rural sanitation coverage through Government Programmes was only 2.53 per cent or 15.87 million. As per the 44th round of the NSS survey (1988-89) 10.96 per cent of the rural population had access to sanitation facilities. The state of sanitation in India is very poor. Sanitation is one of the major indices of the Quality of life. Rural sanitation could not make much headway in the decade. About 611 million rural and 115 million urban people making a total of about 726 million out of 844.32 million (86%) remained unserved at the end of the decade in so far as basic sanitation facilities are concerned.

Targets for the Year 2000

India is committed to the further improvement of water supply and sanitation coverage. Based on the coverage and population forecast, the following target coverages were established.

Rural water supply .... 100 per cent coverage
Rural sanitation .... 7 per cent coverage

The following are the major constraints which still hamper progress in the expansion of service coverage in water supply and sanitation.

Sanitation continues to be accorded low priority. The gap between the provision of drinking water supplies and sanitation facilities is very high. Sanitation coverage is poor because :

- Sanitation is not a felt need
- Infrastructure is inadequate
- Lack of coordination between different implementing agencies
- Insufficient sectoral planning institutional mechanisms
- Inadequately trained human resources Insufficient financial resources and no cost recovery
- Lack of community participation and inadequate health education facilities and
- Lack of private sector participation.

(B) RURAL WATER SUPPLY

Provision of drinking water in the rural area is the responsibility of the State governments and funds were provided in their budget right from the commencement of the First Five Year Plan in 1951. Government of India introduced the Accelerated Rural Water Supply Programme (ARWSP) in 1972-73 to implement the schemes in villages without water supply. The programme continued upto 31 March 1974. In 1974-75 with the introduction of Minimum Needs Programme (MNP), the Accelerated Rural Water Supply Programme (ARWSP) was withdrawn. However, it was reintroduced in 1977-78 when it was found that the progress of supply of safe drinking water to the problem villages was not as per expectation.

In August 1985, the subject of Rural Water Supply and Sanitation was transferred from the Ministry of Urban Development to the Department of Rural Development, with the objective of securing quick implementation and better integration with other rural development programmes. The National Drinking Water Mission was launched as one of the five societal missions in 1986. The Mission was renamed as Rajeev Gandhi National Drinking Water Mission.

A survey was conducted in 1985 to identify problem villages (i.e. without even one source of safe drinking water). In all 1,61,722 problem villages, including spill over problem villages from Sixth Five Year Plan, remained as on Ist April 1986, to be covered in the Seventh Five Year Plan i.e. upto 31st March 1990. Out of these, 1,61,602 problem villages were covered upto March 1996 leaving a balance of 120 problem villages without any source of drinking water facilities. A fresh survey on drinking water facilities in rural habitations was conducted during 1991-94.

As per final results of survey, the position of habitations as on 1, April 1994 is as under :

	Main Habitations	Other Habitations	Total
Not covered (NC)	24113	116862	140975
Partially Covered (PC)	198166	232211	430377
Fully Covered (FC)	337274	410073	747347
Total	559553	759146	1318655

During 1994-95, 70200 Not Covered (NC) and Partially Covered (PC) Villages/habitations were provided safe drinking water facilities and during 1995-96, 81,911 (NC & PC) Villages/habitations were provided safe drinking water facilities. Government has decided to cover 'NC' and 'PC' (getting less than 10 IPCD) habitations by August 1997.

In terms of population, 82 per cent rural population (1991 Census) was benefited by 31 March 1996. The balance population is proposed to be covered during the remaining period of the Eighth Five Year Plan and Ninth Five Year Plan. The total outlay provided for Rural Water Supply during the Eighth Five Year Plan is Rs. 5,100 crore under the Central Sector and Rs. 4954.52 crore under the states/UTs. In 1995-96 the expenditure was Rs. 985 crore under the state Minimum Needs Programme and Rs. 1040 crore was utilised under ARWSP. An outlay of Rs 1.110 crore has been provided under the central sector and Rs 1255 crore under state sector MNP in 1996-97.

Under the existing guidelines the first source of drinking water has to be provided in SC/ST localities and at the time of formulating the scheme, coverage of SC/ST habitations should be given first preference.

From 1990 onwards a minimum of 25 per cent of ARWSP funds have been earmarked for provision of water supply for SCs and 10 per cent of STs. Such funds can not be diverted for only other purpose.

The Government has also set up Rajeev Gandhi National Drinking Water Mission (RGNDWM) authority headed by Prime Minister with concerned Union Ministers, CMs of States, Lt. Governor of UTs as member and Secretary (RD) as member Secretary to review the progress and give directions and policy guidelines for implementation of the programme. In order to implement the directions of the Authority, an empowered committee has been set up with Cabinet secretary as Chairman, concerned Union Secretaries and Secretary incharge of Rural Water Supply in all the States/UTs as member and joint secretary-cummission director as Member Secretary. Similar Authority empowered committee and district committee headed by the Collector have been set up/are being set up in the States/UTs.

(C) CENTRALLY SPONSORED RURAL SANITATION PROGRAMME

According to the National Sample Survey (44th round 1988-89) about 10.96 per cent of rural households has access to sanitary facilities. Taking the figure of NSS as base the coverage has gone up to 13.86 per cent at the end of 1994-95. During 1995-96 target was for construction of 8,70,289 sanitary latrines in the central and state sectors against which the achievement is of 6,89,243 (provisional). In addition to the achievement under Government sponsored programmes the construction of large number of sanitation toilets has been undertaken with private initiative.

The Government has introduced a revised Centrally Sponsored Rural Sanitation Programme since

April 1993, according to which the subsidy is limited to 80 per cent of the unit cost of individual household latrines for persons below the poverty line. The subsidy is shared equally by the Central Government and the State Government. The contribution by the users at the rate of 20 per cent can be provided in the form of cash or kind or labour. The Programme also provides for conversion of dry latrines into sanitary latrines on priority basis.

The total outlay of Rs. 380 crore for the Centrally-Sponsored Rural Sanitation Programme and Rs. 294.23 crore under the state sector Minimum Needs Programme (MNP) has been provided during the Eighth Five Year Plan. The allocation for rural Sanitation programme under the State sector MNP for 1995-96 is Rs. 92 crore while the interim budget has made an allocation of Rs. 60 crore as central assistance to states.

(D) RURAL WATER SUPPLY AND SANITATION IN BIHAR RURAL WATER SUPPLY

1. The State Government has been giving high priority to rural water supply schemes in the Five Year Plans. The progress regarding provision of rural water supply has been quite encouraging but the problem of sanitation exists and needs to be tackled on a massive scale.

2. The U.N. Conference on Human Settlement (HABITAT) decided to declare 1981-90 as International Drinking Water Supply and Sanitation Decade. According to the national goal for the decade, 100 per cent of the rural population was to be provided with safe drinking water and 5 per cent with sanitation facilities.

At the beginning of the Seventh Plan i.e. Ist April 1985 it was assumed that there were 1022 uncovered problem villages (N) with a population of 16.00 lakhs, However under the direction of the Govt. of India a Survey was again conducted which revealed 8177 more problem villages. These were to be fully covered during the Seventh plan period.

Besides these, however there were 13214 partially covered problem villages where water supply needed to be augmented. A survey conducted in the year 1987 revealed 11826 additional partially covered villages. In this way total no. of partially covered villages comes to 26284.

Difficult Areas

In the districts of Dhanbad and Giridih there are certain areas where spot sources of water supply are not feasible. To serve the inhabitants of these area with drinking water supply facilities only course left is piped water supply from surface sources.

It was decided in the meeting of the National Development Council held in June 1990 that there should be no village without a protected source of drinking water. Keeping in view this fact, emphasis has been given on the setting up of the water quality testing laboratories in every district.

From the perusal of the test result of water samples tested by Public Health Institute, Patna, some districts such as Saharsa, Purnea, Araria, Kishanganj, Madhepura, Katihar, Khagaria etc. have excess iron in drinking water which is not good for human consumption. It is, therefore, necessary to set up removal plants in these districts.

Keeping in view the importance of safe drinking water to rural population Rs 40040.00 lakh. was provided in 8th plan. The tribal sub-plan component has been of the order of Rs 13493.00 lakh. and Special Component Plan comes to Rs 14014.00 lakh.

Water Sources

Provision of Rs. 10 crore has been made for completion of on going schemes and standardisation schemes. Rs. 47 lakh has been provided for completion of 250 wells taken up in the year 1991-92. Rs. 371 lakh has been provided for the construction of new wells during the year 1992-97. These wells will be constructed only on the sites where the big machines are not accessible and bore becomes unsuccessful. Out of this amount Rs 30.00 lakh and Rs 263.00 lakh with physical target of 160 and 623 wells is earmarked for sub-plan area and Rs 18.75 lakh and Rs 148.00 lakh for Special Component Plan (SCP) with physical target as 100 and 350 wells.

Rs. 11650 lakh have been provided for the construction of 14424 nos. of handpumps. These include technologically improved types of handpumps such as India Mark III and iron removal plant suitable for hydrological conditions of different area. Out of this amount Rs.115.00 lakh with a physical target of 1735 nos. handpumps is earmarked for sub-plan and Rs. 3713.75 lakh for Special Component Plan with physical target of 45989.

Rs. 11171.00 lakh have been provided for construction of 43884 drilled tube wills. This amount Rs 1619.00 lakh with a physical target of 26395 nos. of drilled tube wells is earmarked for sub-plan area and Rs 3561.50 lakh with a physical target of 13992 for Special Component Plan.

Sanitation

The decadal programme (1981-90) envisages the coverage of 5 per cent rural population with low cost sanitation facilities. The rural population by the end of the decade has been projected to be a round 6.95 crores. 5 per cent of this will be 0.3475 crore. Out of

0.3475 crore, 0.23 crore has been reported covered under different programmes and 0.001 crore has been covered upto 1985-86, 0.0010 crore during 1986-87, 0.0008 crore in 1987-88, 0.000358 crore in 1988-89, nil in 1989-90 and 0.0075 crore in 1990-91. Total coverage is 0.240758 till 1990-91.

The problem of rural sanitation in Bihar is most neglected. Very little has been done in recent years and most village people, live, even today in primitive sanitary conditions. The situation has become somewhat worse ; as some of the good traditions have died and modern facilities have not been introduced to take their place.

Sanitation Programme in Bihar

Minimum Needs Programme (Sanitation Programme)

Year	*Target*		*Achievement*	
	Individual	*Community*	*Individual*	*Community*
1994-95 1995-96	8591	206	177 (2%)	—
1996-97	20012	546	914 (4.5%)	8 (1.4%)

This data highlights that sanitation programme taken up under the Minimum Needs Programme of the Government. This gives the picture of whole Bihar. During the 1994-95 and 1995-96, only 177 individual toilets were constructed which was only 2 per cent of the target. No community toilet was made duirng that period. But during 1996-97 only 4.5 per cent individual latrine and 1.4 per cent community latrine of the target were constructed. These figures reveal a gloomy picture regarding the performance of sanitation programme under Minimum Needs Programme (MNP).

2

REVIEW OF STUDIES

This review of studies is limited to countries of South Asia as the problem of the water supply and sanitation are comparable to those in India.

1. REGIONAL STUDY ON DOMESTIC SHALLOW WELL WATER SUPPLIES

By Kishwar Ijaj, Sultan Ali Barg, and Sazid Hussain Rizvi. (Regional Seminar on Women and Water The Family Handpump, Manila, 1990).

Objectives

The main objective of the study is to explore the extent to which individual families in rural Pakistan can satisfy their basic household water supply needs by self help through family owned wells and the extent to which such a strategy would involve and benefit women.

Methodology

Keeping in view the wide geographic and economic, variations in the country, it was thought appropriate to take the four provinces of Pakistan which would give an adequate idea of the geographic, economic and socio-cultural issues involved. Five villages were selected from each cluster around a focal city. For the selection of villages, the following guidelines were kept in view;

— Year round availability of ground water at depths not exceeding ten metres.

— No piped water supply scheme in the village
— Village population to be 500-2000 persons
— Geographical spread of the village
— Distance from the focal city to be 10-15 km

The research team consisted of a Sociologist and an Engineer who spent one day in each village for the collection of information. The sociologist visited 20 households in each village and interviewed the lady of the house or a relatively senior woman. An effort was made to select households in such a way that they broadly represented the different socio-economic strata in the village.

Findings

It is recognised that much of the population in villages is overlooked in the government planning programmes. Of households surveyed, 60 per cent had their own private wells, 10 per cent used public wells and the rest either canal, spring, river or neighbour's well. Only 10 per cent have any sanitation facilities. The shallow family owned well with handpump has some obvious advantages as a strategy to reach the most deprived more rapidly. But in Pakistan geological, geographical and socio-economic factors vary greatly, making it difficult to accept shallow wells as a widespread solution. Cultural factors and language differences also make for difficulties in getting accurate information. In parts of Punjab, the handpump, individually owned, is very popular but not always a source of potable water. It has also created in many locations a severe drainage and population problem for the community. While community leadership for local problem solving is weak, self ownership tendencies are well established in economically developed areas, probably as a direct result of cultural influences, where as in others, the expectation is that the government will take care of the water supply.

The seclusion of women, combined with general indifference to sanitary amenities, is the cause of much hardship. Even in Punjab for women, water was a second priority while latrines were first. In areas of Sind where the scarcity of water was grave enough to have to use animals to cart water over great distances, bathing places and latrine again were not in evidence. The women of the family used a corner of the house with little or no privacy. In other arid areas such as Baluchistan, stored rainwater was used for drinking and as usual, homes had most rudimentary lavatory facilities. It was apparent that the problems in Pakistan appear to be location specific and will have to be tackled on this basis. Women's needs are so disregarded that the inclusion of women in decision-making processes may be very slow to realise, as also the entry of women's organisations into very conservative areas.

In the North-Western Frontier province nobody preferred the communal atmosphere of the village well to a private well. Preference for a private well or handpump was due to the local cultural restrictions placed on the movement of women outside their home. The local leadership was too weak, even non-existent, to inspire people to unite for community projects.

Baluchistan presented a picture of contrast. None of the villages visited had handpumps as their source of domestic water supply. In the five villages under study were private dugwells, public dugwells, Karez (underground flowing water) and rainwater collected in ditches/ponds. In some of the villages water table was very deep. So animal (camel) power was used to draw the water from the well as well as its transportation to the village. Because of the shortage of water there was no drainage problem. Instead of latrines the people of these villages used open fields. People were least conscious about the quality of water except for its taste.

Conclusion

The following is a summary of conclusions of the field survey.

1. The magnitude of water and sanitation problems is different in various parts of rural Pakistan .
2. The technologies in use for water supply and sanitation vary in different areas.
3. Awareness about water sanitation, health prosperity linkage is almost non-existent.
4. Elements of health education are not visible.
5. Drainage and disposal of waste water and solids is a major próblem.
6. Women are playing a major role in almost all parts of the country to provide water in the house.
7. Women's organisations are almost non-existent in rural Pakistan particularly for water supply and sanitation.
8. The domestic water supply and sanitation concept is well accepted and the major reasons were both a real demand and the cultural pattern of rural Pakistan.
9. There is a demand and need for domestic water supply and sanitation systems in rural Pakistan.
10. A substantial part of the rural population living in smaller villages can be served through domestic water supply and sanitation.

Specific Recommendations

It was recommended that :

1. Rural population should be made continually aware of the importance of clean drinking water and sanitation.

2. Since this project addresses the needs of smaller villages adequate planning system should be developed by the government to include other relevant issue such as sanitation and health education. Awareness of health related to water and sanitation needs to be raised for community motivation.
3. Water supply and sanitation projects should be integrated with the other income-generating activities promoting community participation.
4. Women's co-operatives which manage credit and subsidy facilities should be promoted.
5. The existing infrastructure of women's and youth's organisations for water supply and sanitation projects should be reinforced.
6. This is a family option or a private sector concern. However, for those who can not afford the costs, all possible sources should be tapped to attract financial support from institutions like Zakat fund, people's works programme, national, international, statutory and voluntary organisations.
7. Provision should be made for training of local artisans and mechanics to maintain and repair the handpumps and other sanitation facilities.
8. A research and development programme should improve upon the present quality of handpumps.
9. Government should take measures to develop rural water supply and sanitation standard for Pakistan.
10. Laws and rules should be made to protect aquifers from pollution.
11. Provision should be made for periodic water quality inspections.
12. Facilities like latrines should be integrated in the scheme and a marketing facility be provided

for systems that are best suited to the particular rural community.

2. THE BANGLADESH COUNTRY PAPER

By Tajkera Khair and A.S.M. Imadadual Islam. "Regional Study on Domestic shallow well water supplies" Regional seminar "Women and Water. The Family Handpump" Manila,1990.

Objectives

The objective was to explore the extent to which individual families in developing countries can satisfy their basic water supply needs by self help through family owned shallow well and the extent to which such strategies would involve and benefit women.

Methodology

Before collection of data from the selected villages, questionnaires were presented in the Bhaluka Upzila of Mymensingh district.

The representative villages were to be selected on the basis of some criteria set by the Bank for data collection. It was very difficult to select the villages matching the criteria because the information required was not available at the national level. Therefore, firstly, districts and Upzilas were identified at national level on the basis of secondary data collected from UNICEF Master Plan of Operations (MPO), Bangladesh Water Development Board (BWDB), Department of Public Health Engineering (DPHE) and other related government organisations secondly the representative villages were selected by adopting a random sampling.

Several types of questionnaires were used for collecting data and information from the villages and involved Government and non-Government organisation (NGOs) and institutions.

The methods adopted during data collection are listed below:

- interviewing village committees,
- holding public meetings in the villages,
- holding discussions and meeting with the personnel of concerned government, NGOs and other Institutions,
- interviewing heads of households in the villages,
- personal observations in the villages.

Findings

The data obtained through the survey of the villages under study reveal that the health education and health care facilities in the area are similar to those available to the rest of rural Bangladesh. The majority of the families do not have the basic necessities of life like food, safe drinking water, sanitary systems or even a livable house. Thus the families suffer constantly from diseases and the status of child health is far below acceptable levels.

An analysis of the results shows that the population is badly affected by diseases like diarrhoea, dysentery, worms, fever, malaria, skin disease, measles, etc. Goiter and leprosy and malnutrition complicate the scene. Most of these diseases are carried by water or are due to the absence of acceptable sanitation systems. The death rate on account of these diseases is quite high among children where the rate has gone up, as is evident from national statistics. The present child mortality rate in Bangladesh is 13.1 per cent in the age group of 1-4 years.

Government expenditure on rural water supply and sanitation amounts to no more than about $ 5 million per year. UNICEF/DPHE records indicate about

10 per cent of the rural population use some 7,00,000 public wells and the remainder use other sources such as streams, ponds, springs, etc. Bangladesh has high seasonal variation in the water table, thus wells are not always usable. Many shallow aquifers run the risk of pollution both inorganic and organic, from poor hygiene. Therefore, an integrated water supply and sanitation project would be appropriate.

A source of water near, or better still in the home, would reduce the work burden and improve family health. Generally people have to cover 100 to 200 metre distance to fetch water and in some instances, three to four km. Water salinity is high in some areas, but still acceptable for purposes of washing or bathing. In other areas, rainwater is stored in inhygienic Jars for up to six months as a sweet water alternative to the undrinkable saline water from dugwells. In the last decade however, there has been a massive exploitation of aquifers through shallow irrigation tubewells, as a result of which the water table has been lowered and underground resources depleted. Thus in some areas, particularly in the north, dugwells with handpumps may not now be so popular or feasible. Of the households surveyed 24 per cent used public wells, 16 per cent had their own wells and 60 per cent used pond, river, canal or a neighbour's well. Only 2 per cent of households had adequate sanitation facilities. About 37 per cent used the temporary 'hanging' latrine and the remainder used no facilities at all.

Bangladesh rural development is well supported by rural banks in the formal and co-operative sectors'. The study found that often the less affluent villagers were willing to pay more for a handpump than households which had greater assurance of water. Credit facilities were acceptable with certain conditions, i.e. no interest charged, no collateral required and there should be a three to four years period of repayment, at not less than TK 30 per month.

Income generating activities seem to be more attractive than an improvement of water supply alone. Water supply strategies, particularly that of the family handpump, should, therefore, be linked to training, technology transfer, and teaching entrepreneurial skills. It was felt that sanitation, drainage and health education are also important supporting components of a handpump project.

In the discussion, doubts were raised about the real participation of women in the survey, It was explained that households are indeed dominated by men, but special efforts were made to include women's opinions. This was done by employing various strategies such as timing meetings to suit women.

3. HYGIENE EDUCATION IN BANGLADESH

By Marieka Boot, UNICEF, New York ,1995. This case study was prepared through 20 days of interviews and documentation research and 8 days of writing at the International Water and Sanitation Centre.

Objectives

The main objective was to cover both hygiene education and social mobilization experience and actions in Bangladesh. This was to describe the national rural water supply and sanitation programme and its process of change from pure technical implementation of water supply improvements to a broad movement for improved sanitation and health through hygiene education and social mobilization. The second objective was to collect basic background information about the population, health and hygiene education in Bangladesh. The third objective was to present an overview of hygiene education activities in Bangladesh.

Methodology

Information was collected through individual and group discussions with government and project staff. NGOs, donor organisations and consultants, local authorities and men and women in villages.

Findings

It is estimated that there are some 2.5 million tubewell handpumps in the rural areas out of which one-third are government and two-third are private tubewells. A recent national survey (Mitra, 1992) showed that 85 per cent of the rural population and 98 per cent of the households in urban slums and fringes have access to tubewell water within 150 meters. Accessibility to tubewell water increases with socio-economic status. About 90 per cent of the public tubewells are working. Of the private tubewells, 96 per cent are functional.

Improvements in sanitation are lagging behind improvements in water supply. According to this survey sanitary latrine coverage was estimated at 6 per cent in 1987, rising to 10 per cent in 1989 and to 26 per cent in 1991. The estimate is 33 per cent coverage for 1993. But this means that 67 per cent of the rural population either are using unsanitary latrines or practicing open defecation. About 60 per cent of the sanitary latrines are of the homemade type, while the remainder are water seal latrines. In the urban areas, 48 per cent of the population have access to sanitary latrines.

Various studies have been carried out in support of the water, sanitation and hygiene education activities in Bangladesh. Some of these studies included aspects related to hygiene behaviour and beliefs. A brief summary of their findings is given below.

The national survey by Mitra (1992) indicated that 96 per cent of the population use tubewell water for drinking. But despite relatively general access to tubewell water, only 16 per cent of the households in rural areas and 55 per cent in urban areas use this safe water for all personal and domestic needs. Other main water sources are ponds and rivers. Other studies confirm the picture of a general and exclusive use of tubewell water for drinking, while water from ponds and rivers is preferred for cooking dish washing, bathing, clothes washing and other domestic purpose (Kamal and Choudhary, 1993), (Batman, 1993) Reasons cited for not using tubewell water for purposes other than drinking include : tube-well is too far away (68%) tube-well water turns cooked items black and is distasteful, 'boiling requires more time and spots utensils' (32%) and washing utensils in the pond is more convenient (22%) (Kamal and Choudhary, 1993).

The national survey by Mitra also indicated that although over 90 per cent of families with a sanitary latrine use it regularly, only 10 per cent of children under five and 50 per cent of the older children use the latrine. A recent WHO study found that 25 per cent of the latrines were used by all members of the family (WHO, 1993). Latrine use is higher and more regular among women and girls than men and boys. Also people in urban slums and fringes are more likely to use the latrine than people in rural areas. The needs assessment by Kamal and Choudhary reports that 24 per cent of men and women do not use the same household latrine. Usually an infant defecates in a bed, napkin, cradle or its mothers lap. In most cases, napkins and cloths are cleaned in ponds, canals and rivers. People having little or no access to these water sources wash these clothes with tubewells water, mostly on the platform. Children up to four years of age defecate on the verandah or in the courtyard or open space near the house. Sometimes small pits are specially made for them.

Faeces are cleaned with straw and disposed of in the canal, river or latrine. Faeces are also cleared away with spades and disposed of in ditches. In many cases, faeces remain at the defecation site.

Hand washing after defecation and before preparing or taking food is a general practice, but is not always followed, especially by children. Hand washing is done only with water, soap, mud soil or ash. The survey by Mitra found that the frequency of hand washing with only water is very low after defecation (<5%) and very high before serving or taking food (>95%). The explanation seems to be that after defecation a lot of people use mud for hand washing, whereas before touching food nobody uses mud. Most mothers used water washed both hands and rubbed hands at least three times. Fewer mothers used soap, ash or mud or used a clean rag or air-dried their hands. Several studies show that knowledge about the relation among water, sanitation, hygiene and health is generally low, although most people are aware that drinking unsafe water can cause diarrhoea. Use of sanitary latrine and washing hands with soap or ash are not often mentioned as measures for protecting health. There is a general belief that worm infestation is caused by eating sugar. Knowledge about diarrhoea and the means to prevent diarrhoea is low. Some "22 per cent of the mothers interviewed could not name any cause of diarrhoea, while 36 per cent gave at least one of five traditional causes such as bad air, evil eye, breastmilk, indigestion and hot temperature. Only 3 per cent of the mothers could name more than two `medically correct' causes of diarrhoea. Most commonly mentioned by far were the food hygiene-related causes. Almost three quarters of the mothers could not name even one means of preventing diarrhoea. Those that could name a means of preventing diarrhoea, generally knew few

Kamal and Choudhary are a bit more positive about people's understanding of disease transmission. Some 76 per cent of the indepth interview respondents described a sanitary latrine as a place which flies/ mosquitoes/poultry can not spread bacteria or 'environment is not polluted'. Reasons for diarrhoeal diseases were mentioned to be; eating rotten contaminated food 'drinking polluted water and using polluted water for cooking' flies on food', open defecation, not washing hands after defecation and before taking food.

Mitra's study reports that some 7 per cent of the rural population and 11 per cent of the urban population had received some Health information related to water, sanitation and hygiene in the previous three months. The main sources of information were the health system (health and family planning workers, doctors) and the school system (teachers, male and female students). Neighbours and relatives are also important information providers.

Striking Features

The general assumption supported by studies and personal communication, is that health and hygiene education is not a priority in the Ministry of Health and Family Welfare. There is, in fact, no direction from higher level to carry out hygiene education and many members of the field level staff simply do not consider it as one of their duties. It is little wonder, then, that so few field level staff feel committed to carrying out their educational tasks. Their attempts to do so would, in any case face such obstacles as a lack of initial training (especially in communication skills) refresher courses and in-service training or supervision and monitoring to improve performance. In addition, both field allowances and provisions for transportation to work sites are insufficient.

A Study by the Bangladesh Rural Advancement Committee (BRAC, 1990) provides a vivid account of the thana level obstacles faced by Health programmes, including those in hygiene education. For example, it was found that Health Assistants (HAs) do not devote even one hour in a week to Health education. The report notes, "What is more frustrating is the way health education is delivered. What Health Assistants (HAs) delivered was very superficial and too often given in a very haphazard way. They never tried to make mothers participate: It was one way communication. HAs did not seem to care whether mothers learned. In turn, the mothers were not interested in what the HAs were saying.

4. EVALUATION OF PAST RURAL LATRINE PROGRAMMES IN INDIA, TECHNOLOGICAL ADVISORY GROUP (INDIA) 1985

A number of programmes for the construction of latrines in rural India were undertaken by various agencies—Government, International, non-government etc., since the beginning of the First Five Year Plan or even earlier. Some of the programmes succeeded in the sense that the people maintained them. Several others did not probably succeed. In any case the programme of rural latrines in the context of the over all size of the rural sanitation problem and only less than 1 per cent of the country's rural population is expected to have access to sanitary latrines by the end of the Sixth Five Year Plan i.e. March 31, 1985. A number of reasons are advanced for the poor progress, such as the absence of people's participation, absence of inadequacy of effective community involvement and motivational support, weaknesses of the institutions charged with the responsibility for implementation of the programme, inappropriate or unaffordable technology inadequate dissemination of information on the health benefits of latrines and many others. While reasons are many, they need a fresh look and analysis based on study and evidence to draw lessons for the planning of a larger programme of rural latrines expected to be launched during the remaining years of the International Drinking Water Supply and Sanitation Decade (1981-1990).

Methodology

The methodology of evaluation was left to the Consultants. The results given by various Consultants are based on different sample sizes and some of the observations are impressionistic.

Findings

The entire issue of rural sanitation should be viewed from social point of view rather than only from the point of technical feasibility. After careful consideration of each study, regional variations and available literature on social aspects such as institutional develo-ment, community involvement with special emphasis on disadvantaged and socially weaker sections, local customs, problems and potentialities etc. A broad methodological framework can be formulated to enlist the maximum possible participation from the community. The identification of resource organisations relevant NGOs and subsequent coordination with them will provide a consistent working rapport for ensuring greater achievements. The training materials need to be reoriented according to need and situation since traditional notions, customs, irrational beliefs and taboos are still prevalent in the rural areas especially with regard to sanitation habit. The entire set of training priority and materials can be revamped accordingly. The promotion of sanitary latrine programme in the rural areas required more social reasoning than technological advice.

Acceptance or non-acceptance of the latrines by the individual or groups along with reason was discussed statewise, since the socio-economic situation of a particular state are not identical to other states.

Gujarat

There is an acceptance of the sanitary programme owing to strong software components, the powerful presence of voluntary agencies as well as support of opinion leaders in favour of the programme. The sanitation programme in Gujarat has historical roots which provided a strong social and institutional base to the present programme. Due to its historical roots, the sanitation activities in Gujarat had taken the shape of a movement rather than a mere programme. The movement for Bhangi Kasta Mukti provided a social awareness among the masses. The Safai Vidyalaya played a crucial role as a link between the people and the government functionaries. The Gujarat experience suggests that effective use of software components (community awareness and participation, Institutional development, Human Resource Development and appropriate technology are pre-requisite to success of rural sanitation programme.

Kerala

The study from Kerala shows that 19 per cent of the houses in the State (Urban and Rural) had waterseal latrines in 1980. The highest percentage was in Ernakulam district (30%) which is a comparatively affluent district in Kerala and the lowest percentage of 11 per cent was in Malappuram district which is an underdeveloped district in Kerala. There has to be a correlation between affluence and sanitation within the state of Kerala. But if we compare the per capita incomes of the various States in India and sanitation coverage, the correlation does not hold good. In the case of rural sanitation however, only 15 per cent of the households had latrines with the highest percentage (23) in Errnakulam and the lowest (9) registered in Malapuram district. This pattern follows the normal premise that the need for sanitation is felt more in urban settlements and, therefore, there is a higher rate of acceptance in urban areas compared to rural areas. The flush toilets are on the increase every year and open air defecation is going down proportionately. It has also been seen that in the panchayats close to the city the rate of acceptance is higher. The Kerala report, however, does not give any indication of the correlation between literacy and the acceptance of latrines.

Maharashtra

The Maharashtra evaluation report which analysed a sample of 106 persons who did not have latrines came out with the following major conclusions :

(a) Most of the respondents do not have latrines because of lack of finance and lack of space.
(b) Most of them use open space for defecation.
(c) Most of them want 75 to 90 per cent of the cost of the latrine to be provided as subsidy for constructing their own latrines.

In the case of people using latrine a sample size of 200 was available for analysis. Of these, 38 have pit type latrines and the others have latrines connected either to septic tank or gas plants or to the pits (which are mostly located outside the compound.) Almost all of them constructed the latrines with motivation from Mahatma Gandhi Smarak Nidhi and through private contractors.

Tamil Nadu

A sample of 100 households was analysed in two villages. 38 per cent of this sample had individual latrines. Out of the people not having individual latrines, 27 per cent were willing to construct their own latrines. Out of the households not having latrines, 26 per cent had no place within their own houses, 6 per cent were living in rented house; 9 per cent were too poor to afford latrines, and 2 per cent preferred open air defecation. In their perception about the need for latrines, 69 per cent of the sample felt that the latrines are necessary in keeping the village environment clean, 13 per cent found it convenient to use it all times while 8 per cent felt control of disease as an important reason and 9 per cent of the households reported avoiding foul smell as the reason for having latrines.

West Bengal

In West Bengal a survey of 50 village latrines was undertaken in 1984. These are individual latrines located within the premises of the households. The average family size was 5-8. The income of the families ranged from Rs. 150 to Rs. 1400 per month. These Villages are close to the sub-divisional town of Dirhatta and therefore, the urban influence was visible. All of them were using pourflush latrines, with pan of fiber glass finish, 98 per cent of the latrines were used fully and 2 per cent partially. Superstructures were present in 52 per cent of the latrines.

Uttar Pradesh

Uttar Pradesh has an interesting comparison to offer between the programmes in central zone and western zone. 157 households in the central zone were surveyed out of whom 90 had access to single pit PRAI type sanitary latrines and 67 had no sanitary latrines.

In one Village, 32 out of 35 householders having latrines were educated. There is a definite finding that literates accept latrines much faster than the illiterate villagers but no conclusion can be drawn that uneducated villagers do not accept latrines. It has been found that they also accept latrines.

5. SANITATION AND CULTURE-I

Social Aspects of Sanitation and Personal Hygiene in a Rural West Bengal region, by Vijai Kochar, 1977.

This paper is based on the field study of Bandipur Anchal of Hooghly district West Bengal in 1968 - 70.

Objectives

The study was aimed at social ecology and social epidemiology of hookworm transmission in collaboration with parasitologists and a social ecologist. The present paper presents selected materials pertaining to hygiene, sanitation and disposal of excreta.

Methodology

A population from tweleve contiguous villages was chosen for the study. A random sample 100 households, (750 persons) was selected for epidemiological survey. A sub-sample of 50 households was drawn from the sample households for detailed anthropological studies. The following kinds of field records have been used for the study :

1. Interview of all sub-sample households members once every month, for 15 months, about the time and place of defecation.
2. Identification of one stool of one subject from each household every month to make direct observations.
3. Observations of the actual sequence of defecation behaviour.
4. Basic census survey of all the sample households.

5. Ethnographic records from well informed persons in the communities and descriptive information presented during quantitave studies.

Findings

In rural Bengali culture, personal hygiene, domestic sanitation, disposal of excreta, interaction with 'polluted' foci are linked with some important culture "Idioms" Most important of these is the idiom of purity and pollution. For a rural Bengali household the social, ethical and ritualistic implications of these habits are integrated with the notions about sacred and profane, sin and virtue, pure and impure, etc. For a Hindu household a certain minimum degree of sacredness is necessary to maintain a normal social status. Human faeces, sputum, filth and the objects directly contaminated with these are considered impure' and defiling.

For a rural Bengali household the ethical ritual and aesthetic overtones of personal conduct are intermingled with the notions about health and disease. A popular text on daily rituals for an orthodox devotee includes procedures, prescriptions and even mantras (sacred chants) to go with cleaning mouth, applying oil, bath, grooming, etc. While sanitation is strongly related to culture, its relation to health has often been found to be problematic. This is well illustrated in studies in Varanasi where the population within the University campus (with modern water supply, sewage system, planned housing, structured landscape and superior living conditions) have practically the same pattern of infections and similar morbidity profile as the villagers near Varanasi. Similar conclusion is found by resurvey in Singur (West Bengal) area.

Studies have indicated that even in the case of specific rural sanitation programmes such as latrines and water supply it is difficult to demonstrate definite gains in health parameters until the innovations have diffused to most of the rural population.

The approach of linking the cost of sanitation programmes to definitive gains in the health as a key decision criteria is not a sound administrative strategy. It overlooks the fact that sanitation in rural milieu, particularly in its early phases is more a matter of ethical, educational, aesthetic and status value than merely a matter of significant epidemiological gains. People are more likely to perceive cultural values associated with sanitary innovations than its health implications. Improvements devised in terms of health benefits can be made more acceptable if the technology and its mode of delivery is designed in terms of perceived cultural values. This may be advisable even if in the process some proportion of health benefits is lost. A largely acceptable but a sub-standard device may have better pay off than a 'standard' device with very poor acceptance. Subsequently it may be far easy to replace a substandard device with a device of superior standard.

Sanitation and hygiene is more than any thing else, a way of life. Innovations and technology selected for rural sanitation must be so selected, adapted and presented as to be a part of the existing rural life style and environment in terms of their perceived values and priorities and their existing pattern of habits and customs. This can be achieved by incorporating epidemiologically useful components of the local culture into sanitation programmes. Only then grassroot technology and design can be conceived by the professionals who are concerned about absolutely safe standards, high technology and cosmopolitan values.

6. SANITATION AND CULTURE-II

Behavioural Aspects of disposal of excreta in a Rural West Bengal Region. by Vijai Kochar, 1977.

This research paper is the outcome of the field study in Bandipur Anchal of Hooghly district, West Bengal in 1968 for about two years.

Objectives

The main objective of this paper is to examine this socio-epidemiological strategy with regard to hookworm infection in relation to habits and patterns of disposal of excreta.

Methodology

Method is mentioned in the earlier reviewed first part of the paper Sanitation and Culture 1 Social aspects of sanitation and personal hygiene in a Rural West Bengal region.

Findings

People defecate at short distance from their homes. About 95 per cent of the subjects choose their defecation spot within a walking distance of three minutes from their homes. The males on the average go a longer distance than the females. About 72 per cent of them select a socially recognised defecation ground. They walk around within the defecation ground for an average of 1.4 minutes to reach a suitable spot for squatting. About 69 per cent of the subjects choose a maximally polluted area. They squat for about 3 minutes on an average. About 17 per cent complete defecation within one minute and 71.5 per cent within four minutes. Time lapse between defecation and entering a pond for ablution is less than one minute for 41 per cent of the subjects, less than four minutes for 83 per cent and less than five minutes for 90 per cent.

Only 9 per cent of the households reported having some kind of a latrine and of these households only a few members actually use the latrines. Only 0.8 per cent of the sub sample passed stools in latrines. The people in general consider latrines as dirty, bother some, difficult to service, unaesthetic and uncomfortable.

7. RITUAL PURITY AND POLLUTION IN RELATION TO DOMESTIC SANITATION

By R.S.Khare, 1962. Eastern Anthropologist, vol.15(2). This paper is based on the field study of Gopalpur, a peasant village in Uttar Pradesh. Data was collected during 1958-60.

Objectoves

The main objective of the research paper is to establish that a relationship exists between the ritual status of purity and pollution and the physical states of cleanliness and uncleanliness in domestic surroundings. This paper attempts to point out the nature of ritual purity and pollution involved in the domestic activities and also underlines the importance of studying it with public health aims in view.

Methodology

Methods have not been discussed in the research paper although the author has taken a multicaste village with a population of 572 people.

Findings

Within a homestead, certain places and things are dirty but not polluting or certain things are dirty and polluting but the members of the household do not come into frequent contact with these, and pollution through them can, therefore, be easily avoided. The value of cleanliness is always culturally defined and changing with time. It may on occasions correspond to the modern notions of cleanliness owing to the vigorous emphasis on modernism.

Villager's version of domestic sanitation in a large measure overlaps with the practice of ritual purity and pollution. There are areas that promote or prevent or preclude sanitation. The two beliefs systems, that is the public health one and rural one, contribute towards maintaining 'good' health. Both are the 'means' to good health, but the definition of what constitutes good health is difficult. Since the two systems have overlapping situations, we may expect that there will be contact areas of particular situations wherein the understanding of the two systems will be similar or convergent, either for specific activity or in the structure of the logical interpretation. Such points of similarity, convergence, or correspondence constitute linkages or transference points between the two systems. If the public health workers' discourse in the conflict areas could be re-structured so as to conform to familiar notions or the pattern of logic acceptable to the villagers in the linkage areas, presumably a common basic understanding would emerge, which will increase communication effectiveness and efficiency. The discovery of Pseudo-scientific practices, may thus make it easier to bring in, a phased scheme, the scientific standards of cleanliness in the society.

Studies in Bihar

1. STUDY OF RURAL DRINKING WATER IN BIHAR (INTERIM REPORT)

By K.K. Verma,A.N. Institute of Social Studies, Patna, 1981, (Mimeo).

This study was conducted in the A.N. Sinha Institute of Social Studies and it was sponsored by Rural Electrification Corporation. The aim was to assess the magnitude and nature of the problem of convenient and safe water supply in Bihar for drinking and other domestic purposes. The study included secondary sources at the State level and field observations at the micro level. Eight districts, three from the central Bihar plane, three from north Bihar and two from Chotanagpur and Santhal Pargana. From each district two villages were chosen from among the functional safe water supply scheme villages, one in which there was good performance and the other where the performance was poor. Thirty respondents were chosen from each of the two selected villages. Besides these 30 respondents were also chosen from the fringe settlements of the above schemes. The fringe region was defined as that lying within a radius of 1-3 kms from the safe water supply villages. Some case studies were also taken up. A number of officials involved in the planning and management of rural water supply also interviewed.

The report contains the villagers' perception of the problems the water consumption for drinking and domestic purposes and the management aspects.

Although water sources are there they are not working due to lack of minor repairs and the ones that are functional are not properly maintained. Thus owing to inadequate supply of potable water, traditional sources of water like ponds, tanks, streams and open wells are being used to meet the water requirements in rural areas. These sources neither assure adequate supply of water nor are they always available all the year round. Piped water supply is almost negligible. The few that do exist, are accessible to a selected group of people who can exercise their influence on the local bureaucracy. The most widely used source of water supply is the handpump but its maintenance is poor. Platforms have not been constructed around their installation as a result of which waterlogging is a perennial problem which is detrimental to general health.

Despite the avowed promises and proclamations of arranging water supply for the scheduled caste and scheduled tribe communities, location of the source of water supply stands out as a serious problem which causes inconvenience to these communities. Thus wrong choice of location of water sources, which has been primarily guided by socio-economic and political factors, lack of awareness among the scheduled castes and scheduled tribes of their rights to own a source within accessible distance have been the two important reasons for poor and adequate water supply.

Sugestions

Considering the frequent power failure, the rural water supply scheme should have overhead reservoir. The water should be pumped up in the reservoir whenever power supply is available and then supply be made as per time schedule. At each pump house, two operators should be posted. As it is difficult to manage with one, another should be there. There should be some Advisory Committee at local level to suggest the ways and means for optimal use of the system. The committee may consist of local level functionaries. village people and an official of the Bihar State Water Sewerage Board should be involved in the work. The committee may also be charged with the responsibility of vigilance. There should be a committee at the district level comprising officials from R.E. division of State Electricity Board, District administration, Bihar State Water and Sewarage Board and Panchayat Samities to achieve proper coordination between different functionaries, to expedite the work of rural drinking water supply.

2. 'DHAMDAHA' : A CASE STUDY

By P.M.Tripathi in *L.C.Jain (ed.) Grass without Roots, Sage, Delhi 1985.*

This study was conducted in Dhamdaha block in purnia district. He confirms the findings of other scholars about the poor maintenance of water supply particularly handpumps. He has attributed this to the persistent official apathy and indifference. One fifth of 437 handpumps were found to be defunct.

3.RURAL WATER SUPPLY: AN EVALUATIVE STUDY

By N.Mohsin, A.N.Sinha Institute of Social Studies, Patna Mimeo (1988).

This survey was carried out to examine the status of water supply in the villages of two blocks each selected from 12 districts of Bihar. The study found the maintenance of water sources specially the handpumps badly neglected. The performance of PHED, the nodal organization for planning and execution of rural water supply has been quite dismal. The maintenance wing is ineffective.

3

RESEARCH DESIGN

The problem of rural water supply and sanitation is most neglected. Very little has been done in recent times and most village people live even today, in primitive sanitary conditions. The situation has become somewhat worse. While some of the good traditions have died, no modern facilities have been introduced to take their place. The entire issue of rural sanitation should be viewed from social point of view rather than only from the point of technical feasibility.

Objectives of the Study

1. To assess the existing situation of water supply, public hygiene, habits and sanitation in rural areas of Bihar.
2. To study the role of channels of communication, opinion leaders, mass media etc. in the spread of the schemes of rural sanitation in Bihar.
3. To examine the role of variables like, sex, education, caste, religion, income, space, exposure to media with respect to sanitation.
4. To find out the problems in the adoption of latrine in individual households.
5. To explore the impact of habits, attitudes and beliefs on personal hygiene and sanitation in rural areas.
6. To bring out the views of the youth on the future pattern of rural sanitation.

Methodology

A longitudinal study has been conducted along with situation analysis of water supply and sanitation in rural Bihar. The universe of the study, sampling design, tools of data collection and its analysis are detailed below—

(a) Universe of the study

The universe of the study is Bihar. Since Indian rural society varies from region to region within a state due to diverse geographical, economic, historical, ethnic and other factors; different cultural zones have been covered such as Maithili, Bhojpuri, Magahi and tribal.

(b) Sampling design

There are 13 divisions in the four regions of Bihar. A cluster of ten villages from each division have been selected on the basis of random purposive sampling. Four segments of the village population have been interviewed on the basis of age i.e. grandfather, father, son and women. This has covered three generations of men and women in the whole village. 520 group interviews have been conducted in the all 130 villages. These group interviews covered all the sections of the village population. The perception of the community regarding sanitation and rural water supply has been explored through group discussion with the help of an interview guide. The socio-cultural factors and obstacles faced by the villagers regarding sanitation was also discussed.

(c) Tools of Data Collection

Primary Data: The primary data regarding this study have been collected through the schedule, interview guide, village profile and observation notes. Schedules used for data collection were two, the village schedule and household schedule. The village schedule elicits information on the socio-economic situation in the village as well as status of latrines, drainage of water and problems regarding sanitation and water supply faced by the villages. The household schedule brings out information regarding socio-economic condition, sanitation as well as rural water supply and hygiene behaviour. Information has been collected through frank and informal conversations with the opinion leaders, villagers, women and official staff.

Secondary Data: Secondary data have been collected from available literature such as official reports and other published materials about sanitation and rural water supply in the different cultural zones. Published materials regarding hygiene behaviour and defecation practices in different cultural zones were not available.

Village Schedule : Through the village schedule the collection of information regarding population, number of households, description of educational institutions in the village, number of literate people and mode of transport was done. The sources of water, number and status of latrines in the village as well as condition of drains in the village were also obtained by this schedule. The major thrust of information through village schedule was the problem of water and sanitation in the villages. So far as sanitation is concerned, data on latrine, garbage, and household waste, rubbish, etc. was collected. These were expected to reveal the constraints regarding solution of problems of water supply and sanitation. The village schedule also highlights the diseases due to contaminated water and the community attitudes. The views of the community regarding participation and use of community property resources were also gathered.

Household Information Schedule: Household Information Schedule was used for the identification of head of the household, sickness profile, details of water sources and location of latrines in the individual households. The description of the use of latrine by women and children was the main focus. Waste water disposal, use of cattle dung, details about the sources of information about the hygiene and sanitation, nature of information required and the most preferred sources of information were also enquried from the respondents through the house-holds information schedule. The constraints behind the construction of toilets in the houses were also broughtout.

Interview Guide for Group Discussion

There were some points for the group discussion in the interview guide for different groups of the community like youth, the aged and women. Through the interview guide the main intention of the study was to explore the thinking of the different groups. Through the interview guide the opinion of the group regarding hygiene, sanitation and drinking water was gathered. The social, religious and ethical aspects of the sanitation was elicited from the people . The main focus was on cultural constraints in the adoption of latrine in the individual households. The gender dimension of sanitation was also discussed.

Observation

Observation points reveal the actual situation of community sanitation, drinking water, their source and up-keep, condition of drains, garbage disposal, disposal of human excreta and habits of villagers regarding cleanliness and storage of drinking water. Researchers paid attention to these points while they were collecting field data. This imparted insight to the investigators regarding the sanitation and water supply in the village without asking anybody. The whole picture of the village emerged after observing the situation in the different tolas/hamlets of different castes. Sanitation differs among different castes in the village.

Sanitation is not considered as a health problem among the villagers. During the field work investigators had to devote more time to find out the problems of health. Generally people deny the existence of any problem of health. But depth interviews revealed that there is incidence of diarrhoea, cholera, dysentry, gastroentritis, skin diseases, cold, kalazar, malaria and other diseases. They could not express their views regarding health and sanitation. Although they were much more concerned with the availability of water, they were least concerned about the quality of water.

Field Experience

Researchers went to the field for data collection for the situation analysis of water supply and sanitation in rural areas. First of all researchers went to the specified district and block. Researchers met the Block Development Officer (BDO) in most of the cases. The BDO advised on the selection of cluster villages and contact persons. In most cases *Mukhia* was the main resource person for the field. Rapport was established with the villagers with the help of *Mukhia*. Investigators explained the objective of the study to the villagers. The villagers did not understand the purpose of this research. They thought that Government investigators had been sent for the construction of the latrine. The investigators explained clearly that they belonged to Sulabh Institute of Development Studies, Patna. But villagers presumed that they were Government staff. They thought that the names noted by the investigators will be submitted to government. People were more interested in subsidy and construction of latrines, they were least concerned with the objectives of reserch. They were expecting that their houses would have latrines at government cost. Most of them were least concerned with sanitation.

Group Interviews

Group interview is an open discussion in a small group of people on a specific subject. In the discussion, the emphasis is on a free exchange of views and experiences. The discussion may be recorded on tape. The interviewer's role is to act as a facilitator, stimulating the participants with new points.

During the group discussion with men, women and youth, investigators found that one or two persons were monopolising the discussion. The group as a whole was not participating. This was the case with women as well as youth group. This was due to lack of education as well as exposure to outside contact.

In short, sometimes investigators found the group indifferent due to lack of any action scheme for the villagers. They were more interested in enlisting for the free construction of latrine in their houses. Majority of the group member from the poorest section of the community put up demand.for housing as well as handpumps. They were least conscious about latrines or personal health and hygiene.

Villager's Expectations

Villagers expectations were very high. They demand hospital, English medium schools and more handpumps. They complained regarding the school teachers who took least interest in teaching and showed more interest in party politics. Teachers usually reach the school at noon. Villagers were very unhappy with the government school. In the same way they complained that middlemen who are connected with M.P. or M.L.A. take money for the handpumps. Some villagers suggested sanitation campaign for achieving better health. Villagers' opinion regarding Sulabh Sauchalay or any other flush latrine is that it is very costly . They do not know the cost of construction. This is why many people told that it will cost Rs. 15,000.00 to 20,000.00. This is also the reason for their indifference regarding latrine construction. The poor and landless were much more concerned with the house and handpumps.

The Socio-Economic Setting of the Research

Bihar is the second largest State of India. It covers 5.3 per cent of the area of the country but accounts for 10.23 per cent of its population. The Scheduled Castes and Scheduled Tribes form 23 per cent of the entire population of the State. Nearly 75 per cent of them are below poverty line and suffer exploitation, discrimination and oppression. The situation of women of poor families, in particular, is one of the worst in the country. With high maternal mortality, very low levels of literacy their participation in public affairs is poor. They have least access to skills and inputs and often suffer from an overpowering fear, release from which is essential for any socio-economic and cultural change.

According to 1991 census the population of Bihar was 86,338,853 as against 84,393,0861 of the country. The rate of literacy in 1991 was 38.54 per cent against all India level of 52.11 per cent. Density of population has grown from 402 in 1981 to 497 persons per square km in 1991. About 87 per cent people in Bihar live in rural areas against 76 per cent in the country.

Map. 1 : Map of Bihar

In the context of a critical health scenario, Bihar has been hailed at the national level for its exemplary work in providing safe drinking water to the people all over the State. It has already achieved the target of providing one tubewell for 250 persons and is on the threshold of a revised norm of one handpump for every 150 persons. One may not be quite sure about the ground reality for want of proper maintenance of thousands of tubewells.

Despite high investment in developing infrastructuctural facilities, existing primary health centres, hospitals and dispensaries have not achieved optimum operational levels. Scarcely 20 per cent of the Health Sub-Centres have their own building.

The Child survival and safe motherhood programme has been launched in 13 districts and control of Diarrhoea campaign has been taken up in 23 districts with Unicef support.

Tribal Zone

In the tribal zone Ranchi, Hazaribagh, Palamu and Dumka districts have been selected for the situational analysis of rural water supply and sanitation in Bihar. Tribal zone has a distinct geographical identity with vast hilly tracts and forests. The zone is predominately agricultural but dotted with industrial units of different sizes here and there. Less than 10 per cent of the cultivable land has irrigation facility and hence the agricultural produce is inadequate to meet the needs of rural population. Naturally, therefore, they fall back upon forest resources to sustain themselves throughout the year.

The main problem of the tribal zone is the acute shortage of drinking water. This is why people drink water from wells, surface springs and rivers. Handpumps are very few in number.

So far as sanitary conditions are concerned majority of the people use open place for defecation. Both men and women go outside for defecation. Hardly 4 per cent people have toilets in their home. But the situation is totally different in Palamu district where the percentage of toilet in our sample, was nil. Men and women go for toilet to nearby rivers, orchards and road side.

Approximately 11 per cent are landless 53 per cent are marginal farmers, 23 per cent are small farmer and 10 per cent are medium farmer and only 1.5 per cent are big farmer in the tribal zone.

In the cluster of 40 sample villages in the tribal zone approximately 53 per cent villages got no benefit from health centres.

Bhojpuri

Rohtas, Saran and Muzaffarpur have been included in this cultural zone.

Muzaffarpur is located north of the Ganges. 82.57 per cent people are engaged in agriculture and allied activities. Muzaffarpur district is flood prone. There is large scale migration of agricultural labourers to places like Punjab, Haryana, West Uttar Pradesh and Assam.

Rohtas is most prosperous in the agricultural production, having large canal facilities. Wells and handpumps are the main sources of drinking water. The number of handpumps has increased significantly during last five years.

Rohtas is one of crime infested districts in the western part of Bihar. Most of the schools were without any toilet facility in the school campus.

Magahi

Aurangabad, Jamui and Bhagalpur have been selected as the sample districts for the study purpose.

Aurangabad is the Naxalite dominated district in Magahi zone. Jamui is drought prone district in Bihar. Drinking water is the main problem in the area. People still face the problem of drinking and irrigation water. In some villages people collect water by digging holes in the river bed. This acute shortage prevented people from thinking about the construction of toilets in their houses.

Ten villages of Gidhour block of Jamui district have been surveyed for the situational analysis purposes. Even inhabitants of remote villages knew the name of 'Sulabh Shauchalaya'. People in Gidhour suffer from acute shortage of water.

Maithili

Madhubani, Purnea and Supaul districts have been selected for the study. All these districts are flood affected. People face problem during rainy season due to lack of toilets. Filth and dung are found near the houses. Garbage and faeces are the common sight in the villages. Due to sanitation problem and lack of safe drinking water people are afflicted by diseases like diarrhoea, goitre, malaria, influenza and kalazar.

The survey work was conducted in ten sample villages of Madhubani, Purnea and Supaul districts.

As far as sanitary and rural water supply situation is concerned the bulk of the people defecate in the open. Mostly handpumps do not have a platform, pucca drain and soak pits. Consequently there is waterlogging and water spreads out in streets and lanes. People are still not aware of cheap toilet. They always think that toilet is a very costly affair and common men can not afford it. However, some people want to construct private toilets in their houses.

4

PROFILE OF RESPONDENTS

In the last chapter the research design has been discussed. In this chapter the profile of the respondents is being highlighted through the field data. The religion, caste, community, education, occupation and income of the respondents have been brought out. From this the socio-economic profile of the respondents emerges. The facilities regarding water use, rural sanitation as well as hygiene have been related to the socio-economic status and zone wise distribution of the respondents. Percentage is given in the parenthesis of the Table 4.1.

Table 4.1 : Distribution of Respondents religionwise.

Sl.No.	*Zone*	*Hindu*	*Muslim*	*Others*	*Total*
1.	Tribal	390 (97.5%)	—	10 (2.5%)	400
2.	Magahi	246 (82%)	51 (17%)	3 (1%)	300
3.	Bhojpuri	273 (91%)	27 (9%)	—	300
4.	Maithili	273 (91%)	23 (7.6%)	4 (1.3%)	300
	All	1182 (90.9%)	101 (7.7%)	17 (1.3%)	1300

Altogether above 91 per cent the respondents are Hindu, nearly 7.7 per cent are Muslim and 1.3 per cent Christian, 18 per cent Scheduled Tribes, 17 per cent Scheduled Castes and 39 per cent other Backward Caste and approximately 25 per cent Upper Caste is the percentage distribution of caste and community in all the cultural zones of Bihar (Table 4.2). The scheduled tribes are largely is tribal zone. There is a sprinkling in the other zones. The scheduled caste percentage is highest in Maithili zone and lowest in Bhojpuri zone. Other Backward Castes have a sizeable percentage in all the zones. Similarly the lowest percentage of upper castes has been found in tribal zone.

Table 4.2 : Distribution of Respondents by Caste/ Community

Sl. No.	*Zone*	*Schedule Tribes*	*Schedule Caste*	*Others Backward Caste*	*Upper Caste*	*Total*
1.	Tribal	217 (54.25%)	51 (12.75%)	120 (30%)	12 (3%)	400
2.	Magahi	3 (1%)	99 (33%)	94 (31.3%)	104 (34.6%)	300
3.	Bhojpuri	2 (0.6%)	21 (7%)	175 (58.3%)	102 (34%)	300
4.	Maithili	17 (5.6%)	52 (17.3%)	121 (40%)	110 (36.6%)	300
	All	239 (18.3%)	223 (17%)	510 (39%)	328 (25.8%)	1300

The educational status of heads of the households are brought out in Table 4.3. The percentage of illiterate is highest in tribal zone and lowest in Maithili.

The levels of literacy in different zones were different. Literacy without any level was highest in Magahi and lowest in the tribal zone. The highest percentage up to middle level was found in Bhojpuri zone. The percentage of matric and above matric was highest in maithili zone. It was lowest in tribal zone.

Table 4.4 shows the occupationwise distribution of respondents,

Table 4.3 : Educationwise Distribution of Head of the Households.

Sl. No.	Zone	Illiterate	Literate	Middle	Matric	Post Matr.	Total
1.	Tribal	260 (65%)	32 (8%)	71 (17.75%)	25 (6.2%)	12 (3%)	400
2.	Magahi	108 (37%)	52 (17%)	82 (27%)	33 (10.6%)	25 (7.6%)	300
3.	Bhojpuri	76 (33%)	33 (11%)	91 (30%)	50 (16.7%)	50 (16.7%)	300
4.	Maithili	39 (13%)	36 (12%)	86 (28.67%)	53 (17.7%)	86 (28.7%)	300
	All	483 (37.1%)	153 (11.8%)	330 (25.4%)	161 (12.4%)	173 (13.3%)	1300

Table 4.4 : Distribution of Respondents Occupationwise.

Sl. No.	Zone	Occupation					
		Agriculture	Business	Service	Pvt. Work	Labour & Household	Total
1.	Tribal	352 (87.7%)	7 (1.7%)	16 (4%)	14 (3.5%)	11 (2.7%)	400
2.	Magahi	199 (66.3%)	1 (0.33%)	36 (12%)	60 (20%)	4 (1.3%)	300
3.	Bhojpuri	169 (56.3%)	52 (17.3%)	47 (15.9%)	20 (6.7%)	12 (3.2%)	300
4.	Maithili	218 (72%)	14 (4.7%)	41 (13.7%)	13 (4.3%)	16 (5.8%)	300
	All	936 (71.8%)	74 (5.7%)	140 (10.8%)	107 (8.2%)	43 (3.3%)	1300

The overall percentage of sample household engaged in agriculture was 71.77 in all the cultural zones of Bihar. In the tribal belt percentage of sample households were highest and lowest in the Bhojpuri zone. The percentage of business occupation was highest in Bhojpuri zone while it was lowest in Magahi and tribal zones. In the service occupation the Bhojpuri zone had the highest percentage.

Table 4.5 : Distribution of Respondents Incomewise.

Sl. No.	*Zone*	*Income-per-Annum*					*Total*
		Less than 11000	*110001 to 20000*	*20001 to 30000*	*30001 to 40000*	*More than 40000*	
1.	Tribal	242 (60.5%)	106 (26.5%)	20 (5%)	15 (3.7%)	18 (4.5%)	400
2.	Magahi	196 (65.3%)	96 (32%)	6 (2%)	1 (0.3%)	1 (0.3%)	300
3.	Bhojpuri	175 (58.3%)	43 (14.3%)	63 (21%)	7 (2.33%)	12 (4%)	300
4.	Maithili	110 (36.7%)	49 (16.3%)	59 (19.7%)	20 (6.7%)	62 (20.6%)	300
	All	722 (55.5%)	294 (22.6%)	148 (11.4%)	43 (3.3%)	93 (7.1%)	1300

Table 4.5 brings out the fact that majority of the head of the sample households is below the poverty line. The cut off line of poverty is income below Rs. 11000. In this table we find that 55.45 per cent were below poverty line. But it varied among the different cultural zones.

5

WATER USE, HYGIENE AND RURAL SANITATION

The last chapter dealt with distribution of heads of the sample households. This chapter deals with habits relating to water use, hygiene and sanitation in different cultural zones of rural Bihar. The important reason behind the study of human behaviour is to analyse how successful we are with water use, hygiene behavior and sanitation improvement. Cultural attitudes and beliefs are important motivators for behaviour, but these are not fixed and may be altered because of other changes.

Tribal Zone

Usually people wash their hands with soil after defecation. To take bath is not essential in the tribal belt. People take bath once in a weak. They take fresh water during winter and rainy season. Houses of the tribal people are neat and clean. There is no arrangement of drain in the most of villages. Women sweep their houses regularly and throw garbage in the nearby bush or fields.

The situation of government wells is pitiable. During rainy season surface water goes into the well. The platform of the most of the government wells is broken. The same is the condition with the handpumps and chuan (surface springs). There is no effort for community management of wells, handpumps and the chuan.

Non-tribals of the tribal belt take regular bath while Christian tribals do not forget to take bath on

Sunday. Mostly tribal children are in dirty clothes. It was observed that children do not take bath regularly. Mostly tribal people use karanj for bathing and washing their clothes. There is no strict rule of sanitation. Both men and women go outside for defecation.

Majority of the men and women in Palamu district face acute shortage of water. This is why they take their meal without taking their bath. This situation persists over a long period of time. When they do not get adequate water for drinking, water is not used for sanitation purpose. This was the response of the villagers. People were compelled to drink dirty water from well or chuan (Surface spring).

Magahi Zone

The situation in villages differed according to their distance from the river, in regard to sanitation. The streets in villages near the river were dry while the streets of villages far from river were dirty and water logged. There was difference in the consumption of water also. The water level is lower in the villages close to the river. There are not many handpumps in the villages. So they have to drink water from the well.

But this was not the situation in distant villages. The consumption of water was more. Consequently streets were more dirty. Most of the villages had pond/ tank/*pokhar* which were being used by the livestock.

Whether it is small or big there was filth in every village. Open fields were used by the small children and women for the toilet purpose. Usually people throw their garbage in the streets. Very few people threw garbage at a fixed place so that it was used as compost.

There is no arrangement to clean the village streets. Streets in the village were cleaned at the time of marriage. People in densely populated village demand toilets as there is no place for defecation. Women were facing more problems. Young women of the villages use drains as a toilet. They go without washing their clothes for months. Bad smell comes from the body of poor people. They do not know how to be clean. Villagers clean their house and courtyard mostly on the eve of festivals. But religious women clean and wash their houses with soil or cowdung on Sunday or Tuesday.

Some women clean their hair with clay. Sunday, Monday and Friday are the days when women wash their hair. They feel that hair remains soft and head cool. This is the experience of educated and experienced women. Women do not take bath during menstruation. They take bath after four days. This bath is customary.

It is worth mentioning that women of all castes cook their food without taking their bath. But religious women neither cook nor take food without taking their bath. It is not necessary to take bath during winter. Educated girls and women take their bath regularly except in winter.

Uneducated women do not know that diseases caused by scattered garbage. This is why they throw garbage in the streets and on the road side. Mostly women throw faeces of children on the street or in drains. They do not know that the disease can be caused by garbage and faeces of the children.

In rural areas, people cook their food with dung cake or fuelwood. Due to this they have to live with smoke. Giddiness, eye pain and headache are the common diseases but they do not complain. Poor women and children eat stale food. Even working women have to eat cold food. In some houses people do not cover their food.

Well water is used for the cattle but during summer they use water from the well or river. Mostly wells and handpumps were defunct. During summer handpumps dry up.

Due to lack of toilets small children defecate in the streets and drains. Similarly, women also defecate in drains during the rainy season.

Jamui is more a problem area than other districts. It is much more problematic than Palamu. People have to fetch water from 2 to 3 km during summer. People have to face problems in daily life. They have to dig 3 to 4 feet pit in river for water. Even then people do not get adequate water.

People throw garbage anywhere. Mostly women throw garbage in the streets. Drains are choked with garbage. There is no arrangement for cleanliness. Working women particularly scheduled castes do not give importance to bath and cleanliness.

During summer people take bath after two three days; after digging a hole in the river bed. Cleanliness differs from caste to caste and on the basis of education and religion. Working women from backward and Dalit castes were least concerned with bath and cleanliness. Higher caste women who do not work outside take bath regularly. They have time to pay attention to personal hygiene.

During rainy season there were more mosquitoes due to water logging. People find it difficult to sleep owing to mosquitoes. But during summer season Jamui district has been facing acute shortage of water supply. The reason behind this situation that it is a drought prone area. Moreover, there was inadequate number of handpumps and wells in the sample villages. During summer the situation of villages in Jamui district becomes worse than Palamu.

Maithili Zone

Maithil Brahmins and their houses are more neat and clean than of backward castes in the same village.

Village streets are full of garbage. It is common habit of the villagers to keep dung and household waste in front of their houses.

Mostly, people defecate on the roadside and their fields. Higher castes like Brahmins, Kayastha, Rajput. etc. take bath regularly. Similarly, high caste Muslims also maintain cleanliness. Marwaris are very neat and clean. On the other hand Telis put on dirty clothes. The life style and houses of backward and Dalit castes were dirty. There was waterlogging near the water source. Mismanaged drains is the common characteristic of the villages. In most of the villages drains were *kachcha*. Due to lack of toilets, it is common habit to defecate near by or in field, garden and bandh. Educated people keep water in bucket or pitcher with cover while uneducated did not. Most of the backward and Dalit castes and their children are dirty. The hamlets of dalits are more dirty. Human excreta was scattered on the approaches to villages. Poultry birds eat nightsoil and scatter it on the road.

Tribals and their houses in the Purnea district were neat and clean. There was no garbage near the tribal home. Women and children, utensils and clothes were neat and clean.

Due to the concern for purity Maithil Brahmins have good community surrounding and houses are comparatively clean. Brahmins take regular bath and then perform worship. It promotes environmental health in the village.

Very few Yadav, Halwai, Baniya and Chamar were neat and clean. The children of backward and Dalits were also dirty. The houses and surroundings of the backward and dalit were also dirty. Nightsoil, household waste and garbage were put on the road near the houses. Hamlets of Dom were more dirty due to their rearing of pigs.

Villagers drink water from private and government wells and handpumps. Tanks are used for bath washing and drinking water for cattle. Christian tribal and *refugee* Bangali maintain cleanliness. Home and courtyard of tribals were more clean than those of higher castes of the same village.

Those people who do not have toilet use bamboo groves, road and field for defecation.

Bhojpuri Zone

The state of sanitation in the Bhojpur region was poor. The situation of big villages was worse. There is no idea regarding rural sanitation. The villages and hamlets of *Harijan* castes were more dirty. Conditions are slightly better in the high caste street, but there is lack of cleanliness near the source of water. The surroundings of wells and handpumps are full of slush. Villagers have to defecate by the road side near by the villages. Girls and women use drains as toilet.

Dalit people drink water from well in Harijan tola where garbage and rubbish are heaped. People were compelled to drink dirty water. This is why there were more cases of diarrhoea during rainy season. Dalit women are habituated to live in dirty surrounding. They do not pay attention to cleanliness. The same situation exists in the streets of Muslim dominated area. Generally women do not cover their food. Working women have no time to pay no attention to the cleanliness of their children.

Maintenance of water sources is a more serious problem. There is no provision for maintenance of wells and handpumps by the community.

Cleanliness is not a habit of rural people. As people are poor there is low standard of living in the village. People live in dirty condition. Even the bed sheets were very dirty. The tank or pond in the village were used for washing and bathing purposes. People defecate near the pond and use it for ablution. Night soil is scattered near the pond and around the handpumps.

(A) WATER USE PATTERN IN BIHAR

The source of water for drinking and cooking is mentioned in the Table 5.1 and 5.2. The existing situation of disposal of dirty water in streets, drains and back of the house, open pit and open places is mentioned in the Table 5.3.

Table 5.1 : Distribution of Respondents according to Source of Water for Drinking purposes.

Sl. No.	*Source of Water*	*Zone*				
		Tribal	*Magahi*	*Bhojpuri*	*Maithili*	*All*
1.	Personal & Neighbour's Handpump	24 (6)	166 (55)	237 (79)	249 (83)	676 (52)
2.	Govt. Handpump	11 (2.8)	14 (4.3)	49 (16.3)	49 (16.3)	122 (9.4)
3.	Dugwell	351 (87.8)	119 (39)	13 (4.3)	2 (0.7)	485 (37.3)
4.	River	0 (0.0)	1 (0.3)	0 (0)	0 (0.0)	1 (0.08)
5.	Ponds	0 0 (0.0)	0 (0)	0 (0)	0 (0.0)	(0.0)
6.	Chuan	13 (3.3)	0 (0)	0 (0)	0 (0)	13 (1.0)
7.	Others	1 (2)	0 (0)	0 (0)	0 (0)	1 (0.9)
	Total	400	300	300	300	1300

Note : The figures in Brackets indicate percentages.

Table 5.1 deals with water source for drinking purposes in all the four cultural zones of Bihar. 1300 households were surveyed from different castes, community, religion, occupation for this purposes. This table deals with source wise analysis of drinking water. The main sources for drinking purposes are personal handpump, government handpump, well, river and *chuan*.

Personal and neighbourhood handpumps were found in 6 per cent households in tribal zone, 55.33 per cent in Magahi zone, 79 per cent in Bhojpuri and 83 per cent in Maithili cultural zone. The percentage was 52 in all the cultural zones taken together.

As far as government handpump is concerned, it was 2.75 per cent in Tribal, 4.33 per cent in Magahi, 16.33 per cent in Bhojpuri and Maithili cultural zones. The percentage of dugwells for drinking purposes in 87.75 per cent in the tribal zone, 39.67 per cent in Magahi, 4.33 per cent in Bhojpur and 0.67 per cent in the Mathili zone. Chuan and streams were also used for drinking purposes in the tribal zone. In Magahi zone river water was also used for drinking purposes particularly in Jamui district which is drought prone area in Bihar. There was acute shortage of water for drinking purposes. This is why people go to river and dig pits for the collection of water. In Palamu and Ranchi district contaminated water from *chuan* were also used for drinking purposes. The dirty water goes to *chuan* and it becomes unhygenic.

Table 5.1 shows that majority of the sample households uses its own or neighbour's handpumps. The percentage of users of Government handpumps was about 10 per cent sample households. People also use well for the drinking purposes. In the tribal belt wells are used for the drinking purposes as it was revealed by the 87.7 per cent sample households. It is more in Palamu district i.e. 90 per cent. Eighty per cent sample households in Ranchi district use wells for drinking purposes. It was found that river water is also used for the drinking purposes. In Palamu district *chuan* was also used for drinking purposes as reported by 10 per cent sample households.

Table 5.2 : Distribution of Respondents according to Source of Water for Cooking.

Sl. No.	Zone	Handpump		Well	River	Ponds	Chuan	Other	Total
		Personal	Govt.						
1.	Tribal	19 (4.8)	5 (1.3)	369 (92.3)	1 (0.3)	1 (0.3)	4 (1.0)	1 (0.3)	400
2.	Magahi	121 (40.3)	6 (2.0)	173 (57.3)	—	—	—	—	300
3.	Bhojpuri	237 (78.8)	49 (16.3)	13 (4.3)	1 (0.3)	—	—	—	300
4.	Maithili	245 (81.7)	47 (15.7)	2 (0.7)	4 (1.3)	2 (0.7)	—	—	300
	All	622 (47.8)	107 (8.2)	664 (51)	6 (0.5)	3 (0.2)	4 (0.3)	1 (0.07)	1300

Note : The figures in brackets indicate percentages.

Table 5.3 : Distribution of Respondents according to Disposal of Dirty Water.

Sl. No.	*Zone*	*In Street*	*In Back side*	*Sock-pit*	*In Open pit*	*In the drain*	*Drain at open place*	*Water dry at open place*	*Other*
1.	Tribal	74 (17.9)	79 (19)	0 (0)	11 (2.6)	3 (0.7)	50 (12)	185 (49.9)	10 (2.4)
2.	Magahi	175 (55)	47 (14.7)	1 (0.3)	8 (2.5)	3 (0.9)	73 (22.9)	8 (2.5)	3 (0.9)
3.	Bhojpuri	28 (6.9)	64 (15.8)	3 (3.9)	104 (25)	7 (1.7)	136 (33.6)	16 (3.9)	46 (11.3)
4.	Maithili	11 (2.3)	125 (27)	3 (0.6)	208 (45.2)	7 (1.5)	64 (13.9)	15 (3.2)	27 (5.8)
	All	288 (18)	315 (19.7)	7 (0.4)	331 (20.7)	20 (1.2)	323 (20)	224 (14)	86 (5.3)

Note : The figures in brackets indicate percentages.

In tribal zone as a whole 3.25 per cent sample households use *chuan* for the drinking purpose.

Table 5.2 presents the zonewise analysis of source of water for food purpose in the rural Bihar. Approximately 48 per cent sample household use own handpump. The percentage is highest among Maithili zone where more than 80 per cent use their own handpump. River water is used for cooking meals. Although this situation prevails in Magahi zone particularly in Jamui district where river water is used for cooking during summer season. In tribal zone water from chuan, pond and river is used for the cooking meals.

Table 5.3 reveals that very few sample households use soak pit for the disposal for dirty water in the villages. In the tribal zone soakpit is not found. Drains are found in less than 1 per cent of the sample households. Majority of the sample households reported that dirty water drained out either in the street or dried at open places. In Magahi zone 55 per cent respondents reported that dirty water had gone in the street. But in Bhojpuri zone in 33.6 per cent households water drained out. In the Maithili zone 45 per cent sample households used open pit for waste water. Altogether 20 per cent sample households used open pit for the disposal of waste water.

(B) RURAL SANITATION IN BIHAR

The existing situation of disposal of faeces of children and women, requirement of toilets, disposal of garbage is brought out in the Tables 5.4 to 5.6.

Table 5.4 shows the distribution of households according to the disposal of nightsoil of children below 5 Years in all the cultural zones of Bihar. In tribal zone scarcely 3 per cent households dispose of the nightsoil in the toilet while 97 per cent people use bari, field and orchard for the disposing such nightsoil.

But the situation is different in other cultural zones like Magahi, Bhojpuri and Maithili where 12 per cent, 16 per cent and 16.6 per cent people use toilets respectively for the purposes. More than 11 per cent people use toilets for the disposal of faeces. In Magahi zone more than 86 per cent sample households reported the place of disposal of latrine was bari, field, orchard and river. In Bhojpuri zone 81 per cent households use bari and field for the disposal of nightsoil. But in Maithili area approximately 86 per cent sample households use bari, orchards, field and river.

Table 5.4 : Distribution of Respondents According to the Disposal of Faeces of Children below 5 Years.

Sl. No. *Zone*	*Place of Disposal of Nightsoil*						
	Toilet	*Bari*	*Bari & Field*	*Bari & Orchard & field*	*River*	*Other*	*Total*
1.Tribal	11	66	259	64	—	—	400
	(2.8)	(16.5)	(64.3)	(16)	—	—	
2. Magahi	38	24	233	—	—	4	300
	(12.6)	(8)	(77.6)	(0.3)	(1.3)		
3. Bhojpuri	48	154	92	—	—	6	300
	(16)	(51)	(30)	—	—	(2)	
4. Maithili	50	131	62	48	9	—	300
	(16.6)	45.3)	20.6)	(16)	(3)	—	—
All	147	375	646	112	10	10	1300
	(11.3)	(28.8)	(49.6)	(8.6)	(0.76)	(0.76)	

Note : The figures in brackets indicate percentages.

In all the zones toilet is used by the 11.6 per cent sample households for the disposal of nightsoil of children below 5 years. But majority of the households are utilising open places like bari (Kitchen garden), orchard and river for such disposal. This behaviour

Table 5.5 : Percentage Distribution of Respondents According to Place of Defecation of Women above 14 years

Sl. No.	Zone	Place of Defecation						Total
		Toilet	*Nr.the house Bari*	*Bari Field*	*Bari Orchard*	*River*	*Other*	
1.	Tribal	11	5	247	123	11	3	400
		(2.80)	(1.3)	(61.8)	(30.8)	(2.8)	(0.3)	
2.	Magahi	39	1	251	2	7	0	300
		(12.8)	(0.3)	(84.0)	(0.7)	(2.3)	(0.0)	
3.	Bhojpuri	83	2	77	107	29	2	300
		(27.8)	(0.7)	(25.7)	(35.7)	(9.7)	(0.3)	
4.	Maithili	58	7	47	167	21	0	300
		(19.3)	(2.3)	(15.7)	(55.7)	(7.0)	(0.0)	
	All	191	15	622	399	68	5	1300
		(14.6)	(1.1)	(47.9)	(30.6)	(5.23)	(0.3)	

Note : The figures in brackets indicate percentages.

also depends upon the availability of water. Jamui where there is lack of water, only 2 per cent sample households dispose of the nightsoil in the toilet.

This table 5.5 shows that place of defecation widely used is bari, field and orchard majority of the households in all the cultural zones like Tribal, Magahi, Bhojpuri and Maithili go to the open field for defecation.

Women above 14 years from 14 per cent sample households use toilets. The rest still use open field for defecation. The percentage of toilets is lowest in the tribal zone compared to all other zones. River banks are also used as toilet by the 5 per cent households in all the zones.

Table 5.6 : Percentage Distribution of Respondents According to the Requirement of Toilet.

Sl. No.	*Zone*	*Requirement Percentage*
1.	Tribal	46
2.	Magahi	37
3.	Bhojpuri	65
4.	Maithili	80
	All	56

This table shows that altogether 56 per cent sample households feel the need of toilets in their houses. The percentage is lowest in Magahi zone i.e. 37 but in the Maithili zone 80 per cent sample households voiced the need for toilets in their houses.

In Magahi zone, Jamui and Bhagalpur districts 20 per cent and 4 per cent respectively felt the need for toilet. In Aurangabad about 86 per cent sample households required toilets, while in the tribal zone altogether 46 per cent felt the need of toilets in their houses. In the sample households about 34 per cent in Palamu, 36 per cent in Dumka, 55 per cent in Ranchi and 62 per cent in Hazaribagh felt the need of toilets.

(C) EXISTING SITUATION OF HYGIENE BEHAVIOUR AND DISEASES IN BIHAR

The existing situation regarding disposal of garbage is mentioned in the Table 5.7. Without latrine, lack of safe drinking water and without safe refuse disposal, it is very difficult to prevent the spread of germs. The prevalent diseases in the different cultural zones are mentioned in the table 5.8. Diarrhoea, cholera and malaria were the main diseases after cough, cold and fever.

Table 5.7 : Distribution of Disposal of Garbage according to place.

Sl. No.	Zone	In Pit	In Street	Compost Pit	Hither & thither	In the Field	Other Place
1.	Tribal	86	92	6	122	76	18
		(21.5)	(23)	(1.5)	(30.5)	(19)	(4.5)
2.	Magahi	43	92	10	124	72	12
		(10.2)	(37.5)	(2.3)	(29.6)	(17.2)	(2.8)
3.	Bhojpuri	79	31	26	114	45	77
		(21.2)	(8.3)	(6.9)	(30.6)	(12)	(20.6)
4.	Maithili	231	8	18	26	47	29
		(64.3)	(2.2)	(5)	(7.2)	(13)	(8)
	All	439	223	60	386	240	136
		(28.3)	(18.5)	(3.8)	(24.9)	(15.4)	(8.7)

Note : The figures in Brackets indicate percentages.

Table 5.7 reveals that disposal of garbage is done by very few people at a proper place. 1.3 per cent households tribal 2.3 per cent in Magahi, 6.9 per cent in Bhojpuri and 5 per cent in Maithili zone used compost pit for disposal of garbage. Altogether 3.8 per cent sample households use compost pit for the disposal of garbage. This shows that people have no idea about sanitation. Even street and open places were used for the garbage disposal. Overall 15 per cent sample households kept garbage in the agricultural field. Altogether 80 per cent sample households were least concerned about proper disposal of garbage.

No compost pit was found in the Dumka and Palamu districts in the tribal zone. Most sample households in Palamu district kept garbage in their agricultural field. But in the Dumka most of the sample households kept garbage either in the pit, street or open places.

Table 5.8 : Distribution of Respondents by Diseases.

Sl. No.	*Zone*	*Cold- Cough, fever*	*Diarhoea*	*Malaria*	*Kalazar*	*Cholera*	*Polio*	*Others*
1.	Tribal	32	29	89	1	3	2	8
		(19.5)	(17.6)	(54.2)	(0.6)	(1.8)	(1.2)	(4.8)
2.	Magahi	14	13	38	15	21	1	—
		(13.7)	(12.7)	(37.2)	(14.7)	(20.5)	(0.9)	—
3.	Bhojpuri	201	11	8	3	12	—	2
		(84.8)	(4.6)	(3.3)	(0.4)	(5)	—	(0.8)
4.	Maithili	288	5	2	—	2	—	1
		(96)	(1.6)	(0.6)	—	(0.6)	—	(0.3)
	All	535	58	137	19	38	3	11
		(66.7)	(7.2)	(17)	(2.3)	(4.7)	(0.3)	(1.3)

Note : The figures in Brackets indicate percentages.

Table 5.8 shows that members of the households have been victim of diarrhoea, cholera, and malaria during the past one year as reported by the respondents of tribal as well as Magahi zone. But in Bhojpuri and Maithili zones people reported main diseases like cough, cold and fever. Altogether 7.2 per cent sample households reported that members of the family have been victims of diarrhoea while 17 per cent suffered from malaria and 4.7 per cent from cholera. Malaria was more found in Ranchi, Hazaribagh and Dumka.

6

SOCIAL VARIABLES AND THE EXISTING SITUATION

This chapter deals with the social variables in relation to the existing situation in the rural areas of Bihar. The distribution of households with own toilet facilities and water facilities, caste, community, religion, income and occupationwise are discussed in different tables. The reason for not having toilets, attitude towards community toilets and women sanitary complex is also analysed in this chapter. Role of channels of communication regarding toilet are also described towards the end of this chapter.

All these matters are discussed in the tables given below :

(A) WATER FACILITY

Life without water is inconceivable. Potable water is the essential prerequisite for health. In the absence of piped water supply people in rural areas have to depend on sources of water such as ponds, tanks, streams, open springs and wells. Sometimes in summer even these sources dry up. The greatest hardship is faced by women who have to fetch water for household purposes from a distance and by scheduled castes who do not own sources of water. It is generally seen that most of the wells and hand pumps are located in that part of the village in which they do not have access.

Table 6.1 : Religionwise Distribution of Respondents having Water Facility.

No. of HHs.	*Tribal*	*Magahi*	*Bhojpuri*	*Maithili*	*All*
Hindu	390	246	273	273	1182
Total	(97.5)	(82)	(91)	(91)	—
Water	96	115	210	233	654
Facility	(24.6)	(46.7)	(76.9)	(85.3)	(55.3)
Muslim	0	51	27	23	101
Total	(0)	(17)	(9)	(7.6)	(7.7)
Water	0	35	25	22	82
Facility	(0)	(68.6)	(92.5)	(95.6)	(81)
Others	10	3	0	4	17
Total	(1.5)	(1)	(0)	(1.3)	(1.3)
Water	3	0	0	4	7
Facility	(30)	(0)	(0)	(100)	(41)
Total	400	300	300	300	1300
Water	99	150	235	259	743
Facility	(24.7)	(50)	(78.3)	(86.3)	(57)

Note : The figures in brackets indicate percentages.

The percentage of sample Hindu households having water facility was 24.6 per cent. On the whole 24.75 per cent sample households were having water facility in tribal cultural zone. In the Magahi zone about 47 per cent Hindu, 69 per cent Muslim were having water facility. Altogether 50 per cent sample households were having water facility. Similarly 78.33 per cent sample households had own water facility. This facility is higher in the Maithili cultural zone. But overall percentage of all the cultural zones is about 57 per cent sample households with own water facility. The percentage of own water facility is lowest in the tribal zone. This shows that drinking water is still a problem in the tribal zone. The situation is better in the Maithili zone as revealed by Table 6.1.

Table 6.2 shows that 24.75 per cent sample households having own water facility in tribal zone.

The caste and communitywise distribution of households having own water facility were 11.5 per cent among scheduled tribes, 17.6 per cent in scheduled

Table 6.2 : Caste/Communitywise Distribution of Respon-dents having Water Facility.

No. of HHs.	*Tribal*	*Magahi*	*Bhojpuri*	*Maithili*	*All*
STs	217 (54.25)	3 (1)	2 (0.6)	17 (5.6)	239 (18.3)
Water Facility	25 (11.5)	1 (33)	1 (50)	5 (29.4)	32 (13.3)
SCs	51 (12.7)	99 (33)	21 (7)	52 (17.3)	223 (17)
Water Facility	9 (17.6)	25 (25.2)	15 (71)	41 (78.8)	90 (40.3)
OBCs	120 (30)	94 (31.3)	175 (58.3)	121 (40)	510 (39)
Water Facility	57 (47.5)	44 (46.8)	135 (77.1)	113 (93.3)	358 (70.1)
Upper Caste	12 (3)	104 (34.6)	102 (34)	110 (36.6)	328 (25.8)
Water Facility	8 (66.6)	80 (76.9)	84 (82.3)	95 (86.3)	267 (81.4)
Total	400	300	300	300	1300
Water Facility	99 (24.7)	150 (50)	235 (78.3)	259 (86.3)	743 (57.1)

Note : The figures in brackets indicate percentages.

castes, 47.5 per cent in other backward castes and 66.6 per cent in upper castes. Similarly, the percentage of water facility in the Magahi zone was high in the households under the upper castes category that is about 77 per cent and lowest in the scheduled castes that is 25.2 per cent. In the Bhojpuri zone altogether 78.3 per cent sample households were having water facility. But percentage is very high among upper castes.

But situation is better in the Maithili zone where about 86 per cent sample households have water facility. Thc high percentage of own water facility was among other backward castes i.e. 93.3 per cent.

Table 6.3 : Distribution of Respondents having Water Facility by Educational levels.

No. of HHs.	*Tribal*	*Magahi*	*Bhojpuri*	*Maithili*	*All*
Illiterate	260 (65)	108 (37)	76 (33)	39 (13)	483 (37.2)
Water Facility	41 (15.7)	68 (62.9)	49 (64.4)	37 (94.8)	195 (40.3)
Literate	32 (8)	52 (17)	33 (11)	36 (12)	153 (11.7)
Water Facility	6 (18.8)	29 (55.7)	23 (69.6)	32 (88.8)	90 (58.8)
Middle	71 (17.7)	82 (27)	91 (30)	86 (28.6)	330 (25.3)
Water Facility	30 (42.2)	24 (29.2)	71 (78)	17 (81)	285 (86.3)
Matric	25 (6.2)	33 (10.6)	50 (16.6)	53 (17.6)	161 (12.4)
Water Facility	14 (56)	25 (75.7)	44 (88)	50 (94.3)	133 (82.6)
Matric +	12 (3)	25 (7.6)	50 (16.6)	86 (28.6)	173 (13.2)
Water Facility	8 (66.6)	16 (64)	48 (96)	80 (93)	152 (87.8)
Total	400	300	300	300	1300
Water Facility	99 (24.7)	150 (50)	235 (78.3)	259 (86.3)	743 (57)

Note : The figures in brackets indicate percentages.

Majority of the sample households have water facilities. Overall 57 per cent were having own water facility. 40.3 per cent of illiterates who were having water facility. The situation of own water facility were highest in Maithili zone and lowest in Tribal zone. Only 16 per cent illiterates were having water facility in Tribal zone (Table 6.3).

Table 6.4 shows out that sample households engaged in labour and households work do not have own water facility in the Tribal as well as Magahi zones.

Table 6.4 : Occupational Distribution of Respondents having Water Facility.

No. of HHs.	*Tribal*	*Magahi*	*Bhojpuri*	*Maithili*	*All*
Agriculture	352 (88)	199 (66.3)	169 56.3)	218 (72)	936 (71.8)
Water Facility	79 (22.4)	134 (67.3)	130 76.9)	192 (88)	535 (57.1)
Business	7 (1.75)	1 (0.33)	52 17.33)	14 (4.67)	74 (5.69)
Water Facility	2 (28.5)	1 (100)	50 96.1)	12 (85.7)	65 (87.8)
Service	16 (4)	36 (12)	47 (15.7)	41 (13.7)	140 (10.8)
Water Facility	11 (68.7)	3 (8.3)	40 (85)	35 (85)	89 (63.5)
Private Work	14 (3.5)	60 (20)	20 (6.7)	13 (4.3)	107 (8.2)
Water Facility	7 (50)	13 (21.6)	12 (60)	11 (84.6)	43 (40.1)
Labour +	11 (2.8)	4 (1.3)	12 (3.7)	16 (5.7)	43 (3.3)
Water Facility	0 (0)	0 (0)	3 (25)	9 (56.3)	12 (27.9)
Total	400	300	300	300	1300
Water Facility	99 (24.8)	150 (50)	235 (78.3)	259 (86.3)	743 (57)

Note : The figures in brackets indicate percentages.

Moreover, this percentage is lower in Bhojpuri as well as Maithili zones compared to other types of economic activity e.g. Agriculture, Business, Service and Private work. Overall 57 per cent sample households were having own water facility in all the cultural zones taken together.

The percentage of households having own water facilities are higher in the economic categories like service and business where 87.8 per cent and 63.5 per cent sample households respectively have own water facility.

Table 6.5 : Incomewise Distribution of Respondents having Water Facility.

No. of HHs.	*Tribal*	*Magahi*	*Bhojpuri*	*Maithili*	*All*
Up to Rs. 11000	242 (60.5)	196 (65.3)	175 (58.3)	110 (36.6)	722 (55.5)
Toilets	23 (9.5)	60 (30.6)	121 (69)	84 (76.3)	288 (39.8)
Rs. 11001 to 20000	106 (26.5)	96 (32)	43 (14.3)	49 (16.3)	294 (22.6)
Toilets	35 (33)	83 (86)	40 (93)	42 (85.7)	200 (68)
Rs. 20001 to 30000	20 (5)	6 (2)	63 (21)	59 (19.7)	148 (11.4)
Toilets	18 (90)	5 (83)	58 (92)	55 (93)	136 (91)
Rs. 30001 to 40000	15 (3.8)	1 (0.3)	7 (2.3)	20 (6.7)	43 (11.4)
Toilets	10 (66.6)	1 (100)	6 (85.7)	18 (90)	35 (81)
Rs .40000 +	18 (4.5)	1 (0.33)	12 (4)	62 (20.6)	93 (7.15)
Toilets	13 (72)	1 (100)	10 (83.3)	60 (96.7)	84 (90.3)
Total	400 (100)	300 (100)	300 (100)	300 (100)	1300 (100)
Toilets	99 (24.7)	150 (50)	235 (78.3)	259 (86.3)	743 (57)

Note : The figures in brackets indicate percentages.

Table 6.5 shows that higher the income, higher is the ownership water facility in the sample households. But the percentage is more in Bhojpuri and Maithili zones in comparision to tribal and Magahi zones. In the below poverty line category 56 per cent sample households scarcely 40 per cent have own water facility in all the cultural zones. But it is lowest in Tribal zone, where hardly 9.5 per cent have own water facility. In Tribal zone people are facing water scarcity for drinking and other purposes. In all the cultural zones sample households having income level above Rs. 40,000 per year have own water facility.

(B) TOILET FACILITY

Adequate sanitation is the foundation of development but a decent toilet or latrine is an unknown luxury to half the people on earth. On the brink of 21st century half the world's people are enduring a medieval level of sanitation. Million of people do not have access to a decent toilet, and many of them are forced to defecate on the bare ground. Access to safe water and adequate sanitation is the foundation of development. To deny people basic sanitation is not just inhuman, it also kicks the first step out from a country's ladder of development.

Table 6.6 shows zonewise distribution of households with own toilet facilities in rural Bihar. Altogether 14.5 per cent sample households were having own toilet in the house. It ranged between 4 per cent toilet in the Tribal zone to 26 per cent in Bhojpuri zone. The situation of sanitary latrine in Palamu and Dumka was very bad. It is not found in ten cluster villages. This means that cent percent people of Palamu and majority of the people in Dumka were practicising open defecation.

Table 6.6 : Distribution of Respondents with Toilet Facilities.

Sl. No.	*Zone*	*Having Toilets*	*Percentage*	*Total*
1.	Tribal	16	4	400
2.	Magahi	34	11.3	300
3.	Bhojpuri	79	26.3	300
4.	Maithili	57	19	300
	All	186	14.5	1300

The situation of Magahi cultural zone is better than tribal zone where 11.33 per cent sample households have toilets in their houses. But in Jamui district, only 4 per cent sample households have toilets in the house. Water is the problem in Jamui district. It is a drought prone area. This is why people give priority to water over toilet.

In Bhojpuri cultural zone, Muzaffarpur, Saran and Rohtas districts the situation is different. The percentage in Muzaffarpur itself is very high. The reason behind this high concentration is that the sample villages are close to urban and industrial areas like Kanti in Muzaffarpur district.

In Maithili cultural zone Madhubani, Purnea and Supaul districts were selected for the survey 19 per cent sample households had toilet of there own.

Altogether 14.5 per cent sample households in all the four cultural zones had toilets. More than 85 per cent people are still defecating in open place like, road, river bank, fields, orchard, railway track, ponds and forest.

Table 6.7 shows religion-wise distribution of households having toilets in all the four cultural zones of Bihar. Altogether 11.5 per cent Hindu sample households had toilets of their own while the percentage of Muslims was 45.5. When we look at the zone-wise figure of the toilets in the household the percentage is highest in Bhojpuri cultural zone. But it is lowest in Hindu households in that zone. The percentage of others like Sikh and Christian were 23.5 in all the zones taken together.

Table 6.7 : Religionwise Distribution of Respondents having Toilets.

No. of Households	*Zone*				
	Tribal	*Magahi*	*Bhojpuri*	*Maithili*	*All*
Hindu	390	246	273	273	1182
Having Toilets	14	9	64	49	136
	(3.5)	(3.6)	(23.4)	(17.9)	(11.5)
Muslim	—	51	27	23	101
Having Toilets	—	25	15	6	46
		(49.01)	(55.6)	(26.08)	(45.5)
Others	10	3	0	4	17
Having Toilets	2	0	0	2	4
	(10)	(0)	(0)	(50.00)	(23.5)
Total	400	300	300	300	1300
Having Toilets	16	34	79	57	186
	(4.0)	(11.3)	(26.3)	(19.0)	(14.3)

Note : The figures in brackets indicate percentages.

Table 6.8 deals with community and caste-wise distribution of households having toilets in their houses in the rural areas of Bihar. Overall 4 per cent scheduled tribes, 2.6 per cent scheduled castes, 7.8 per cent other backward castes and 41 per cent upper castes have toilets of their own. The lowest percentage of toilets is found in scheduled castes while it is 7.8 per cent in Tribal zone and 3.8 per cent in Maithili zone. But in the Bhojpuri and Maithili zones no latrine

Table 6.8 : Community and Castewise Distribution of Respondents Having Toilets.

No.of Households	*Tribal*	*Magahi*	*Bhojpuri*	*Maithili*	*Total*
STs	217 (54.25)	3 (1)	2 (0.6)	17 (5.6)	239 (18.3)
Having Toilets	8 (3.6)	— —	1 (50)	1 (5.9)	10 (4)
SCs	51 (12.75)	99 (33)	21 (7)	52 (17.3)	223 (17)
Having Toilets	4 (7.8)	— —	— —	2 (3.8)	6 (2.6)
OBCs	120 (30)	94 (31.3)	175 (58.3)	121 (40)	510 (39)
Having Toilets	3 (2.5)	1 (1)	25 (14.28)	11 (9.09)	40 (7.8)
Upper Caste	11 (2.75)	100 (33.3)	102 (34)	110 (36.6)	323 (24.8)
Having Toilets	1 (9)	33 (33)	43 (42.1)	57 (51)	134 (41)
No Caste	1 (0.25)	4 (1.3)	— —	— —	5 (0.3)
Having Toilets	—	—	—	—	—
Total	400	300	300	300	1300
Total having Toilets	16 (4)	34 (11)	79 (26)	57 (19)	185 (14)

Note : The figures in brackets indicate percentages.

was found in the houses of scheduled castes. There are more households having toilets among upper castes. The percentage in Maithili zone is 51 per cent and in Bhojpuri it is 42 per cent. The lowest percentage of toilets in upper castes were found in Tribal cultural zone that is only 9 per cent.

Altogether the percentage of households having toilets of their own is 4 per cent in Tribal, 11.33 per cent in Magahi, 26.33 per cent in Bhojpuri and 19 per cent in Maithili zones.

Table 6.9 highlights the role of education in this respect. As educational level rises the adoption of sanitary toilet goes up. The percentage is lower in those households which are illiterate as compared to the literate and higher levels. Only 11.8 per cent illiterate households had toilets. In the case of households whose head was educated above matric the percentage was 34.6.

Table 6.9 : Educational Status of the Respondents having Toilets

No. of HHs.	*Tribal*	*Magahi*	*Bhojpuri*	*Maithili*	*All*
Illiterate Total	260 (65)	108 (37)	76 (33)	39 (13)	483 (37.15)
Toilets	4 (1.5)	30 (27.7)	21 (27.6)	2 (5.12)	57 (11.80)
Literate Total	32 (8)	52 (17)	33 (71)	36 (12)	153 (11.77)
Toilets	0 (0)	2 (3.8)	3 (9)	1 (2.7)	6 (3.9)
Middle Total	71 (17.7)	82 (27)	91 (30)	86 (28.67)	330 (25.3)
Toilets	4 (5.6)	1 (1.2)	20 (22)	11 (12.8)	36 (10.9)
Matric Total	25 (6.2)	33 (10.6)	50 (16.67)	53 (17.7)	161 (12.4)
Toilets	7 (28)	1 (3)	11 (22)	10 (18.8)	29 (18)
Above Matric Total	12 (3)	25 (7.6)	50 (16.67)	86 (28.7)	173 (13.3)
Toilets	1 (8.3)	4 (16)	22 (44)	33 (38.3)	60 (34.6)
Total	400	300	300	300	1300
Toilets	16 (4)	34 (11.33)	79 (26.33)	57 (19)	186 (14.4)

Note : The figures in brackets indicate percentages.

Maximum toilets are found in those households whose heads had education above matriculation. It was also observed that literates and educated households have more toilets than illiterate.

Table 6.10 reveals the occupationwise distribution of heads of the households having toilets in the sample villages of Bihar.

Table 6.10 : Occupationwise Distribution of Head of Respondents having Toilets

No. of HHs.	*Tribal*	*Magahi*	*Bhojpuri*	*Maithili*	*All*
Agriculture Total	352 (87.8)	199 (66.4)	169 (56.4)	218 (72)	936 (71.8)
Toilets	12 (3.4)	34 (16.5)	43 (25.4)	40 (18.3)	129 (13.7)
Business Total	7 (1.8)	1 (0.3)	52 (17.3)	14 (4.7)	74 (5.7)
Toilets	0 (0)	0 (0)	18 (34.6)	1 (7.1)	19 (25.6)
Service Total	16 (4)	36 (12)	47 (15.7)	41 (13.7)	140 (10.7)
Toilets	4 (25)	0 (0)	15 (32)	14 (34.4)	33 (23.5)
Private Total	14 (3.5)	60 (20)	20 (6.7)	13 (4.3)	107 (8.3)
Toilets	0 (0)	0 (0)	2 (10)	2 (15.3)	4 (3.7)
Labour Total	11 (2.7)	4 (1.3)	12 (3.67)	16 (5.7)	43 (3.3)
Toilets	0 (0)	0 (0)	1 (8.3)	0 (0)	1 (2.3)
Total No.	400	300	300	300	1300
Toilets	16 (4)	34 (11.3)	79 (26.3)	57 (19)	186 (14.3)

Note : The figures in brackets indicate percentages.

The percentage of households engaged in agriculture having toilet was 13.7 per cent. It is higher in case of those engaged in business and service. The percentage is lowest in the category of labour and households work. It is 0 per cent in three cultural zones of Tribal, Magahi and Maithili. Only Bhojpuri zone had 8.33 per cent heads of the households engaged in labour and household work have toilets of their own. Heads of the sample households engaged in private work did not have toilet in their houses while 15.3 per cent in this category had toilets in the Mathili zone. Altogether 3.7 per cent head of the sample households were having toilets in their own houses.

Table 6.11 deals with incomewise distribution of respondents having toilets in the sample villages of

Table 6.11 : Incomewise Distribution of Respondents having Toilets.

No. of HHs.	*Tribal*	*Magahi*	*Bhojpuri*	*Maithili*	*All*
Up to 11000 Total	242 (60.5)	196 (65.3)	175 (58.3)	110 (36.6)	722 (55.5)
Toilets	3 (1.2)	5 (2.5)	30 (17)	10 (9)	48 (6.6)
11001 to 20000 Total	106 (26.5)	96 (32)	43 (14.3)	49 (16.3)	294 (22.6)
Toilets	4 (0.9)	25 (26.04)	18 (41.8)	10 (20.4)	57 (19.3)
20001 to 30000 Total	20 (5)	6 (2)	63 (21)	59 (19.7)	148 (11.4)
Toilets	1 (5)	1 (16.6)	22 (34.9)	19 (32.2)	43 (29)
30001 to 40000 Total	15 (3.75)	1 (0.33)	7 (2.33)	20 (6.67)	43 (11.38)
Toilets	3 (20)	1 (100)	2 (28.5)	5 25)	11 (25.58)
40000 + Total	18 (4.5)	1 (0.33)	12 (4)	62 20.6)	93 (7.15)
Toilets	5 (27.7)	1 (100)	5 (41.6)	15 (24.1)	26 (27.9)
Total	400 (100)	300 100)	300 (100)	300 (100)	1300 (100)
Toilets	16 (4)	34 (11.3)	79 (26.3)	57 (19)	186 (14.3)

Note : The figures in brackets indicate percentages.

Bihar. It indicates that higher income goes with increased availability of toilets. Income plays a vital role in the construction of toilets. Sample respondents below poverty line have low percentage of toilets in the house i.e., only 6.6 per cent in the poverty line category. It is lowest in tribal and Magahi zones. About 28 per cent sample households with the income level of more than Rs. 40,000 had toilets. The situation is different in sample households in the income bracket of Rs. 11001 to 20000. In this category 19.3 per cent households had toilets.

(C) REASONS FOR NOT HAVING TOILETS

The following table highlights the reason for not having toilets in the house. The reasons behind the non-adoption of toilets in the house are also mentioned below :

The reason for not having toilets in the house is mentioned in Table 6.12. The main reason behind this was poverty in the Tribal zone while lack of subsidy was the prime cause for the Magahi zone. In Bhojpuri zone 58.6 per cent household reported the main reason was lack of cheap technology. This was the main factor in the Maithili zone. Overall more people reported lack of subsidy as the main obstacle in the construction of toilets in the house.

Table 6.12 : Distribution of Respondents by Reasons for not having Toilets

	Tribal	*Magahi*	*Bhojpuri*	*Maithili*	*All*
Poverty	129 (32.3)	39 (13)	107 (35.6)	119 (39.6)	394 (30.3)
Lack of Material	7 (1.75)	5 (1.6)	13 (4.3)	84 (28)	109 (8.3)
Lack of Mason	20 (5)	24 (8)	10 (3.3)	16 (5.3)	70 (5.3)
Lack of Water	42 (10.5)	3 (1)	11 (3.6)	14 (4.6)	70 (5.3)

	Tribal	*Magahi*	*Bhojpuri*	*Maithili*	*All*
Lack of Land	9 (2.3)	7 (2.3)	3 (1)	3 (1)	22 (1.6)
Lack of Subsidy	104 (26)	103 (34.3)	166 (55.3)	106 (35.3)	479 (36.8)
Lack of Cheap Technology	79 (19.8)	87 (29)	176 (58.6)	194 (64.6)	536 (41.2)
Bad Thing	1 (0.3)	3 (1)	2 (0.6)	3 (1)	9 (0.6)
Do not Think	6 (1.5)	49 (16.3)	1 (0.3)	9 (3)	65 (5)
Not Required	0 (0)	1 (0.3)	0 (0)	6 (2)	7 (0.5)
Others	4 (1)	8 (2.6)	13 (4.3)	36 (12)	61 (4.6)
Total	400	300	300	300	1300

Note : The figures in brackets indicate percentages.

Overall 30 per cent sample households reported that poverty is the main cause. This is why most people do not have toilet in the house. Lack of land, water and subsidy from the goverment are the major obstacles in the construction of toilets in the houses. Shortage of water is in the Tribal zone was reported by 10 per cent sample house holds as an important reason for not having toilets.

In the Maithili zone, poverty, lack of subsidy and cheap technology were the main reasons behind not having toilets in the house.

(D) OPINION TOWARDS COMMUNITY TOILETS AND WOMEN SANITARY COMPLEX

Table 6.13 reveals that 64 per cent people favoured community toilet while 35 per cent did not. In the tribal zone 56.5 per cent sample households realised the need of community toilet. But 42.2 per cent did not favour the community toilet due to lack of solidarity in maintaining its cleanliness. But in the Magahi zone 58 per cent sample household disfavoured the community toilet. They want individual toilet. The reason behind this was the lack of any arrangement for its maintenance. There was acute dearth of community participation in the water supply and sanitation work. Earlier poeple voluntarily used to clean drain but now these days drains have been root of tension. This sometimes leads to litigation.

Table 6.13 : Distribution of Respondents Favouring of Community Toilets

Zone	*Yes*	*No*	*No Response*	*Total*
Tribal	226 (56.5)	169 (42.2)	5 (1.3)	400
Magahi	125 (41.6)	174 (58)	1 (0.3)	300
Bhojpuri	189 (63)	109 (36)	2 (0.6)	300
Maithili	291 (97)	9 (3)	— —	300
All	831 (63.9)	461 (35.4)	8 (0.6)	1300

Note : The figures in brackets indicate percentages.

The majority of the people from Bhojpuri as well as Maithili zones liked community toilets in their villages.

Table 6.14 reveals that women sanitatry complex is favoured by 57.4 per cent sample households in all the zones. But 42 per cent people disfavored the sanitary complex for the women particularly in tribal and Magahi zones. Majority of the sample households did not favour sanitary complex for women. 52 per cent in tribal, 76 per cent in Magahi zones did not like women sanitary complex. Women preferred squatting outside the village. The main reasons behind dislike of women sanitary complex was the lack of water, lack of community spirit for cleanliness and maintenance of such a complex. Lack of understanding among the women themselves was the root cause for not favoring the sanitary complex.

Table 6.14 : Distribution of Respondents Favouring Women's Sanitary Complex

Zone	*Yes*	*No*	*No Response*	*Total*
Tribal	186 (46.5)	210 (52.5)	4 (1)	400
Magahi	71 (23.6)	228 (76)	1 (0.3)	300
Bhojpuri	193 (64.3)	105 (35)	2 (0.6)	300
Maithili	297 (99)	3 (1)	— —	300
All	747 (57.4)	546 (42)	7 (0.5)	1300

Note : The figures in Brackets indicate percentages.

(E) ROLE OF CHANNELS OF COMMUNICATION

Communication is power, communication is money, communication is fun and entertainment and communication is intelligence today. So a country's growth cultural moorings its inner strength and competitive edge all depend greatly on its communication power.

Communication channel is the course by which a message is conveyed to the receiver. Mass-media and interpersonal channels were analysed in this chapter in terms of the role in the diffusion of innovations whereas interpersonal communication is more likely to cause attitude change. Thus mass and interpersonal communication channels can be complementary rather than competitive in diffusing innovations.

Tradition plays vital role in the formation of ideas regarding sanitation habits and water use pattern. This is the tradition of many parts of rural Bihar that higher castes people do not drink water from the wells used by scheduled castes. Many villagers have already used Sulabh Shauchalaya in their district town during their visit to towns. They came to know regarding sanitary latrines and their uses. Subsequently some of them began to construct toilets in their houses.

Previously wells were used for the drinking purposes but these day due to communication channels people have started using water from handpumps. Handpump water is safer than well water. People pay more attention to the safe drinking water. It helps in reducing diarrhoea, cholera and other water borne diseases.

Most of the respondents reported that the main source of knowledge regarding sanitation was the town visit. The channel of communication plays vital role in awareness generation but people in rural areas preferred inter-personal communication. Mostly households wanted awareness training for the sanitation and use of safe drinking water. People complained that nobody imparts knowledge regarding health, sanitation and safe drinking water. They expressed their inability to listen to radio as they are busy in domestic and agricultural work as well as ignorant about the timing of broadcast. Consequently they prefer inter-personal communication at the time of village meeting.

Majority of the villagers were so poor that they had no T.V. So they were deprived of watching any kind of programme on energy, environment, health, sanitation and safe drinking water.

Similarly newspapers were not easily available in the rural areas. So people remained ignorant of day today developments in the field of health, renewable energy, sanitation and environment.

7

SOCIO-CULTURAL CONSTRAINTS

People are more likely to perceive cultural values associated with sanitary innovations than its health implications. In the hands of technocrats the word sanitation has however, become synonymous with a few technological interventions such as toilets, waste disposal, sewage, water supply, vector control and more recently, air and water pollution control. This has led to the neglect in public health of the whole spectrum of cultural and bahavioural intervention in man-environment relationship which the term 'sanitary' originally signifies.

Sanitation and hygiene as part of health culture is integral part of folk culture (way of Life) or 'Life Style'. Promotion of 'scientific' knowledge, practices, norms, technology, therefore, does not occur in vaccum. The 'scientific' approach itself assumes a way of life (right way of understanding and doing things) which emerges from scientific health culture of cosmopolitan—urban social milieu. Lack of fit between the cultural load of a sanitary device and the competing elements of folk culture creates strong medico-cultural conflicts.

Male Dominance in the Family

Women are the main users and carriers of water but the entire decision regarding sanitation and water supply is taken by the male members of the family. Culturally no decision is final until and unless male members of the family give their opinion. Most of the male members do not consider these things essential. Many women do not take bath regularly due to lack of privacy. Similarly, women do not freely defecate due to the same reason. Sometimes new brides either do not take food or take very light food due to fear of defecation during day. It is a problem particularly in large villages. Mostly poor women do not have their own source of water. They do not clean their body owing to lack of privacy.

Purda System

In the tribal areas there is no *purda* system. Men and women take their bath in the same pond. But in the Maithili, Magahi and Bhojpuri zones, the situation is totally different. Women have to wait to get privacy. They cannot change their clothes after taking bath. The same situation prevails among Muslim and Christian women. In several places women have to wait for dusk for defecation. This creates health and sanitation problems.

Small House Sites

Non-availability of land for the toilets is also a major constraint for the landless, marginal farmers, wage earners and others. Some people want toilets but land becomes the major obstacle. Moreover, villagers do not want toilets inside the house. They consider toilet as polluting. Since houses are small and the kitchen and the toilet cannot be adjacent to each other, the toilet is dropped from consideration.

Folk Practices

Majority of the people do not have their own source of water supply in the rural areas. This is why people prefer to go out for defecation and clean their mouth and take bath in the river. In north Bihar, this is the common practice. They walk some distance from the house to ease themselves and have an opportunity to visit their agricultural field . No extra time is required for this.

In tribal belt there is shortage of water in the house and open land is available for defecation. So most people do not bother about toilet in the house. The reason behind non-adoption is lack of water as well as practice of defecating in the open. So people do not want to change this practice and also put additional financial burden on their shoulders.

Women always prefer to go out for easing in groups. They have no chance of meeting neighbouring housewives. So they get some time to meet and interact with each other. This is the place where they talk about the affairs of each family. They get information not only about the other families but about their relatives in other villages. Similarly, the village well is a place of gossip for young daughters-in-law.

Religion

Men and women devoted to religious rituals do not want toilet in the home due to the presence of ancestral deities. In the same way they do not use water from handpumps for worship. They consider water drawn from the well pure, even though the well may not be covered and dust, dirt and leaves fall into it. People use clay for cleaning their bucket and put the bucket inside the well and draw water from the same well. In this process well water ceases to be clean.

Most people consider drain water as dirty and hesitate to clean the drain near the well or handpumps. This is why the dirty water flows out to the street. Most streets of the village become muddy during rainy season and it is difficult to move through them.

Cow dung is not considered so dirty. It is used at the time of worship. So people hardly mind the heap of cow dung near their house. Many rural households use cow dung cake as fuel. This is why dung heaps are found near the house. They do not mind mosquitoes and other insects attracted by them.

If any disease like cholera, diarrhoea breaks out in the village, villagers attribute it to the anger of the goddess Bhagwati. They do not relate it to polluted water, contaminated food and scattered human excreta in the street. Thus, villagers pay no attention to proper disposal of garbage, cow dung, and faeces of children.

Concept of Purity and Pollution

Purity can be defined as a state of ritual cleanliness, where as cleanliness itself refers to a physical state. In Hindu and Muslim society concepts of cleanliness and filth, purity and impurity are well developed and have a strong effect upon personal and household hygiene. Thus, in some Hindu areas, the wife will only enter the kitchen in a pure state, that is after she has washed herself and changed clothes, and not when she is menstruating (Kochar, 1991). In the Muslim society ritual impurity is the usual state in which one is found. Purification involves washing of one's hands, face and feet before prayer, and taking a complete bath after sexual contact, menstruation, and child birth (Simpson-Hebert, 1984). However, purity and cleanliness are not always two sides of the same coin. For example, a person can observe purity rites and wash hands before prayer but not do this before eating.

Kochar (1978) in his study on hookworm transmission provides us with an example of the relationships between purity and cleanliness. Among rural people in Bengal, notions of the pure and the sacred and of the polluted and profane are in many ways the rules for personal hygiene as well as for ritual. A popular text on daily rituals for orthodox Hindus includes procedures, prescription and even sacred chants to go with cleaning the mouth, applying oil, bathing, grooming and so on.

The canons of 'Folk hygiene' embody some very powerful notions of personal cleanliness.

In many cultures and societies, human excreta are considered to be polluting from the bio-medical model of disease transmission. At the same time, the excreta of babies and little children are often considered to be harmless in all respects. Also, among many people, the left hand is used for anal cleansing and no matter how well this hand is washed afterwards it remains the 'dirty' hand and should never be used for handling and serving food, eating, shaking hands, etc. Hall *et al,* (1991) report in their study on water, sanitation and health that 'the faeces of a child that is only breast fed are not considered to be 'dangerous' (when it has diarrhoea) by the vast majority of people. Only four respondents said 'that infants' faeces were dangerous 'from birth'. The rest offered different ages (ranging from about two to six) but stressed that faeces become dangerous once a child starts to eat solid foods and the faeces begin to smell. This they stressed, was the sign of danger. As for the cause of faeces becoming dangerous, a few interviews suggested the mixing of foods, 'in the stomach, others said that young children's faeces were not dangerous until they have suffered from serious diseases such as tuberculosis'.

The Santhal strongly feel that toilets in the house are polluting and dirty. If it is in the house they have to face anger of the ancestral deities. Such pollution will inhibit the entry of the soul of ancestors. Santhal people are fully concsious of the purity of their house.

Similar situation exists in the Munda region. Old Munda women are conscious of social, religious, economic and cultural obstacles in the construction of toilets in the house. They do not want to infringe their religious norms. They favour the construction of toilets outside the boundary of the village. But young and educated Munda women do not accept these views. Similarly, Christian tribal women do not subscribe to this view. They are ready to give land for the construction of the toilets while unconverted tribal women do not like this idea.

Muslim women feel that if toilet is constructed in the house, its purity will not be affected.

Religious and old Hindu women do not like toilets in the house as it is polluting.

Ritual purity and physical cleanliness are two things; One refers to the ritual or religious state and the other to the physical state. Several words such as *Pavitra-Apavitra, Sudha, Asudha* and *Choot-Pak* are used to devote the ritual states of purity and impurity. Similarly, the words *Saaf Ganda* are used to refer to physical states of cleanliness and dirtiness. The terms *Choot-Pak*, are used in two senses, one in which they refer to situational purity-impurity of an individual 'at a given moment in relation to other individuals, as determined by events occurring in his family or in his own life', and the other in which they refer to the place of a caste group in the ritual hierarchy of caste system. Human beings suffer ritual impurity through natural happenings (birth, death, menstruation etc.) association with material objects (human skin, excreta, menstrual blood, spittle, reproductive fluids, etc.,) and human beings (untouchables, Maha-Brahmins etc.) (Khare, 1962).

Castes and Tribes

Tribal people do not pay due attention to the health and sanitation. They do not have the habit of regular bath. Women take their bath twice in a week.

Christian tribals are more neat and clean. They are more educated. Impact of education on Christians is not seen in the sphere of health and sanitation. The priest plays an important role in deciding any thing including health and sanitation. The hamlets of Dalit castes are more dirty in comparison to other backward and higher castes. The children of Musahar, Dom, Turi etc. are very dirty and pay no attention to personal health and hygiene. Regular bath is exceptional in these children, men and women and working people. They take bath when they have leisure. Although they have money to buy liquor they have no money to purchase soap for cleaning their clothes.

Similarly, houses of Santhal, Munda and Oraon are very clean but the clothes are dirty. They have no idea about dirtiness of the clothes. They are ignorant about personal health and hygiene.

Perception about Transmission of Diseases

The cultural perceptions about causes of water and sanitation related diseases result in behaviours based on the bio-medical perspective. For example, Sinhalese women in Sri Lanka will try to avoid an over consumption of 'hot and heavy' foods during pregnancy and breast feeding, to prevent their babies from getting diarrhoea (Nichter, 1988). Kaltenthaler *et al.* (1988) state that when they asked why hand washing was important as many as 53 per cent of the people mentioned reasons unrelated to the prevention of disease. Gwatirisa (WP 1991) reports that the persons interviewed in her study attached much more importance to the availability of toilets for convenience and privacy reasons, than for reduction of diseases.

Cultural perceptions about causes of water and sanitation related diseases often vary between different groups of people and may change with time.

"We found that very few mothers buried their children's faeces. However, from some very old people, we learned that into the past (and in a very limited way today) mothers 'always' used to bury the faeces of breast feeding infants for a rather peculiar reason. They believed that if a dog ate such faeces the infant would then be inflicted by a serious case of red, foamy diarrhoea. This view is in keeping with more widely held beliefs that a persons' excreta, hair, nail clippings and so on, may be used by another to inflict harm against them. These old people claimed that in their youth i.e. when they were young mothers there was not so much diarrhoea as one finds today. They attributed this to the consistent burial of infants faeces."

A young mother told us that one day she 'tested' the idea by throwing out her child's faeces on to the ash-heap so that a dog could eat them. 'Sure enough, my child soon had very bad red diarrhoea. From then on she was very careful to bury them. This mother was an exception. The vast majority simply discard their children faeces on the ash-heap regardless of the age or health of the child. In other words, old beliefs which inspired (for whatever reason) a sanitary practice are no longer held and the message about burying children's faeces had not been given an equivalent credency' (Hall *et al.*, 1991).

Intra-Familial Relations

In the rural set up socio-cultural factors are more effective than other technological factors as well as health factors. Women do not want to seen going for toilet by senior member of the family. The father-in-law should not know when their daughters-in-law go for toilets. If toilets is located in inside the house everybody will know when their daughters-in-law go for defecation. Women think that this is purely private matter and privacy must be maintained. So Intra-familial relationship is also important factor which affects adversely the popularisation of the sanitation programme. This is why some women do not prefer the construction of toilets in the house.

Women prefer to go for defecation and fetching of water from the handpumps or well in a group. Rural women get an opportunity to communicate or interact with each other at the time of defecation. They talk freely. This is the place or time where daughters in-law complain against their mothers-in-law. They feel relaxed whenever they go for defecation. This is why they never realised the dangers of environmental pollution.

8

VIEWS OF THE YOUTH

No attempt has been made to involve youth in the rural water supply, sanitation and community health programme in Bihar. Water supply and sanitation programme have not been linked with poverty alleviation or income generating programme. The views of the youth are important as they reveal the shape of things in the future.

Rural Water Supply

Tribal Youth

Youths reported that there is acute shortage of drinking water in the village, particularly during summer. Therefore, they have to take water from surface springs. It was also reported that many handpumps were defunct for the last two to three years but nobody took the notice including the government.

Youths of Jarmundi Block of Dumka district revealed that the water problem has been solved through *Jal Hi Jaan Hai* programme. Previously they had to depend on river water throughout the season. Very few youth in the rural areas knew that the contaminated water and faeces of children may lead to spread the disease. Social, religious and economic factors are responsible for the non-adoption of own water supply. Henceforth, the youth wanted for some more information regarding cleanliness of the body, water and village.

Muslim Youth

During rainy seasion some people take water from the government handpump near the mosque. Water from well is used for domestic purposes. Majority of the handpumps are out of order. Government does not maintain them.

Muslim youth desire that there should be a Health guide who can talk about prevention of diseases. During the rainy season water related disease appear but they can be checked through the health awareness. In some of the villages people use hot water during rainy season. There is always scarcity of drinking water. Sometimes people fetch water from nearby villages.

Dalit and Backward Caste Youth

In hamlets where handpumps were not available, people use water from the well. Sometimes drain water goes to the well. Tanks and ponds in the village are used for animal and men, both for bathing and washing their clothes. Children and even old people use water for ablution from the same tank. Villagers defecate near the tank. Surroundings of handpumps and wells are also very dirty. In some areas water from handpumps is not safe. One of our investigators found that water from one handpump was muddy. Headmen of the many villages admitted that clean water is not available in the village. The reason behind this unsafe water, is installation of handpumps through the Jawahar Rajgar Yojana (JRY) scheme. Mostly handpumps have been installed at 60-70 feet depth while under PHED scheme it should be between 100 to 160 feet deep.

Handpumps of M.L.A. quota are given to his henchmen. So poor people are unable to get such facility. Regarding community health, youth reported that headache, gastric, stomach ache, cough, and cold, fever and malaria appear due to filth and mosquitoes. Diarrhoea, cholera are the main diseases in the village and women, children are the victims. Although some people have suffered due to dirt and filth, most of them have developed immunity. However, they want bettter living conditions. No steps have been taken by government in this direction.

Diarrhoea and cholera appear during rainy season due to filth, dung and scattered human excreta in the street. Very few people had compost pit for making manure and keeping the street clean.

Sanitation and Community Health

The importance of toilets in our day to day life is accepted by the educated youth and the villagers. Due to availability of open space in the village, majority of the people do not think about the construction of toilets. But because of education, external exposure and demonstration effect, people have started talking about the construction of toilet. Most youth are still unaware of the development of different types of toilets which provide sanitation as well as manure. Uneducated youth are more concerned with toilets in the home. For them, sanitation means only provision of toilet.

Tribal Youth

In the tribal belt youths do not feel any necessity for toilets as a part of sanitation. The main need is irrigation. Youths were least concerned with the health, hygiene and sanitation aspect of everyday life. They believe that to defecate in the open field is beneficial to health. Their culture and social beliefs do not permit construction of toilets inside the house. There is some flexibility among christian tribal youths. Majority of the youths are uneducated and ignorant and thus they do not know the importance of health and hygiene in life.

Cultural factors are more effective in the decision-making, regarding the adoption of toilets in the house. Many youth reported that it will entirely depend on the permission of Pahan of the village. Although they have been living in the same village but castes and tribes have different views on these issues.

Youths have already used toilets in the town, at least an experience of Sulabh Shauchalaya was found common whenever they visited the nearby district town. Some tribal youths unlike the old women are already aware of toilet necessity, although they feel that it is a costly item and would cost about rupees ten to fifteen thousand. Community toilet is not accepted by youth and old people but at the same time youth preferred loan for the construction of toilets repayable in 6 to 11 months. It appears that provision of subsidy can be a crucial factor as far as expansion of toilet construction is concerned. In the area under study, Sulabh Shauchalaya is also preferred for the twin purposes, its capacity to yield manure as well as its low cost.

Another problem in the rual area is to get trained masons to construct toilets. To defecate in the open field is injurious to health. Any programme for the cleanlines of the villages should be supported through the *shramdan* and donation is the desire of the local youth. Repair of wells and distribution of bleaching powder will improve the quality of water in the rural areas. Even subsidy will not change the attitude of tribal youths because traditionally they have been using open field, river banks and forests for squatting as well as bathing.

They clean their houses but smoke is a problem for them. Youth want some information regarding cleanliness of body, water and village. Santhal youth view toilets as a symbol of modernisation. Its construction inside the house may anger the house gods. Kheturi youth of Dumka district have latrines in the house. But the youth in both communities expect subsidy for the construction of toilets. Both the groups were ready to contribute their labour for any kind of sanitation work in the village.

Both the Kheturi and Santhal youth were in a fix regarding the use of masure which is the by product of Sulabh Shauchalaya. They were hesitant to put this manure in the agricultural field. Educated Santhal youth know the use of latrine while uneducated were not aware of toilet as part of sanitation. Even with subsidy hardly 5—7 families can take initiative for the construction of toilets in their houses. This shows the indifferent attitude of youth regarding the adoption of latrine in the house as well as in the village. The reasons behind the non-adoption of latrine are illiteracy, ignorance, poverty, social and religious belief.

Some youth reported that they have not suffered from any disease due to open defecation. Some youth know about diarrhoea and malaria which are common during summer and rainy season.

They said that private latrines can be constructed without any subsidy while community latrine cannot be constructed without any subsidy. People have realised that latrine is good for health and hygiene. Waterlogging is a grave problem. It creates tension in village. Pucca drains and soakpit near the handpump will improve rural sanitation. Most of the youth know about Sulabh toilet.

Earlier toilets were not considered important, but with the population growth the mentality of the villagers have changed. Now they have realised the need of toilet in the home. Some of the villagers are ready to construct the toilets. But economic and technical constraints are there.

Muslim Youth

Muslim youth prefer individual latrine, but community latrine can be acceptable only in the mosque. Educated women do not want to go outside for the defecation. *Purdah* system is prevalent in the Muslim society. Culture inhibits women to go outside for defecation. Muslim youth do not want to incur a loan for the construction of latrine.

Affluent Muslims can spend money for the construction of toilet but poor Muslims can not afford construction of toilets in their houses. Economic factors are important impediments but ignorance, illiteracy, and lack of exposure are the other factor for non-adoption of toilets.

Dalit and Backward Caste Youth

Dalit and backward caste youth are familiar with the toilets. But they consider it an urban item. Some of these youth in sample villages of Bhagalpur reported that they have already used Sulabh Shauchalaya in Bhagalpur town, Kiul Junction and elsewhere. Lack of land and financial difficulty are the important reasons for not constructing the latrine. Faeces of small children are thrown in the drain, streets and nearby fields. Latrine is not the felt need of the Dalit.

If any NGO or government provides them subsidy for the purpose of latrine construction, they may not utilise the grant properly. They do not realise that the garbage will cause disease. It creates some problem during rainy season. There is the problem of smoke in the kitchen but they expressed their inability to report the disease caused by smoke. Poor and illiterate persons were totally unaware of community health and sanitation.

In villages where drains have not been constructed, dirty water flows in the street. Drain is used as a toilet for children. Heaps of garbage and cow dung pollutes the environment. The streets where the Dalit live were very dirty, even open space near Dalit houses were very dirty due to scattered human excreta and pig dung.

9

WOMEN'S PERSPECTIVE

(A) RURAL WATER SUPPLY

Tribal Women

Tribal women were more concerned about the cleanliness of village, maintenance of well and handpumps and compost pits rather than individual or community latrines. There is lack of water in the tribal area, which affects sanitation and good hygienic behaviour of the rural women. What to talk of bathing and washing of clothes, even safe drinking water is a problem. The main sources of water are surface springs handpumps and wells. But wells are very few in number and the water is not quite clean.

Women and children are the worst sufferers of the acute shortage of water. Women take bath in the river or at the spring or well if it is available.

Due to acute shortage of water, majority of tribal women have to go without bath. Particularly during summer they have to face several problems of health and sanitation. On the one hand they get little water in summer, on the other hand they have to take polluted water from unclean sources during rainy season. Most children are in dirty clothes.

Most tribal women have no handpumps in their houses. Regular bath is a problem for them.

Christian and educated tribal women strive hand to keep neat and clean. Regular bath and washing of clothes are the common feature. Unconverted tribals women are not neat and clean.

Tribal women reported that the irrigation well constructed by the YMCA of Xavier Institute is being used for drinking purposes. Ponds are also used for the bath. In the Ranchi district some ponds have been constructed by the YMCA. Women take their bath in the pond. Water is a problem for tribal women. They complained that most handpumps were defunct in the tribal villages. The condition of government wells is very bad. Dirty water trickles down in the well in the rainy season. Most tribal women in Ranchi district drink boiled water during winter and rainy season. Neither handpump nor surroundings of handpumps are good in the tribal area.

Non-Tribal Women

Upper Caste Women

Safe drinking water is still a problem in the rural areas. Majority of the upper caste women do not understand the importance of safe drinking water. Educated women do so. Water is also a problem. Most wells are in a very bad condition. This is why most women have started using hand pumps for the drinking purposes. Consumption of water from well is being reduced day by day. Majority of the upper caste women have hand pumps or well in the houses or nearby their houses. Some of them are unhygenic as dirt and dust trickle down to the well. Most wells go dry during summer. Even upper caste women take their bath at the well in the open. Only young and educated women draw water and take bath in the house. There is lack of privacy at the well and handpumps.

In the Gidhaur Block of Jamui district people collect water after digging in the river. There was clear cut impact of shortage of water on the people. Women and children take their bath after one or two weeks during summer season. Clothes were not washed regularly.

Owing to lack of water, mostly upper caste women clean their house only on the eve of religious festivals. During summer they get dirty water from the well.

Women are least concerned with the maintenance of well and handpumps. Women are not ready to spend money for the upkeep of community hand-pumps, although they face the problem all the year round.

Other Backward Caste (OBC) Women

Mostly backward caste women from Mallah, Barber, Carpenter, Khan, Dhanik, Koeri, Yadav etc. are more concerned about having handpumps rather than sanitary latrines in their house. They expect hand-pumps from government. Some women of backward castes complained that they had to drink from such wells where drain water trickles down. Lack of privacy in taking bath at the well, pond river is a common problem. Regular bath is not necessary for working women of backward castes. Sometimes bath is taken after a fortnight during winter season. They do not wash their clothes regularly. This is why the children and women of backward castes were more dirty than those of upper castes. Educated, well-to-do and young women of backward castes take bath and wash their clothes regularly.

The adverse impact of shortage of water is clearly visible on the women and children as they do not take bath and wash their clothes regularly. Bad odour comes from their body and clothes. Backward caste women take their bath with dirty and contaminated water. Due to ignorance they use polluted water for drinking purposes.

Backward caste women reported that most government well or hand pumps were in a very bad condition. Even private wells were not maintained. They do not know the importance of sanitation. This is due to illiteracy, ignorance and poverty. Mostly backward women below poverty line are not bothered about the source of water.

Educated, well-to-do and young women were more aware of health and sanitation. Affluent backward caste women have their own source of water. Government wells or handpumps in affluent backward hamlets were in better condition than in Dalit hamlets. They know that contaminated water caused diarrhoea.

Dalit Women

Most Dalit women do not have their own source of water. They have to fetch water from either community handpumps or wells. What to talk of safe drinking water, even water is not available at door step. They have to fetch water for the whole family. Usually they take bath in the river, pond, well or some place out side the house without any privacy. Due to lack of water they have to live in filthy conditions with their children and other family members. Among the Dalit, Musahar, Dom and Turi have no source of water while very few Chamar, Dusadh and Dhobi had own source of water. Educated Dalit women take care to cover water after collection.

Dalit women are more concerned about handpumps than sanitary latrines. They expect handpumps to be installed by the government. Mostly handpumps or well in Dalit tola were in a bad condition. Dirty water was going into the well. The surrounding of handpumps were not good. Drain water was scattered in the street. Dalit women have to use polluted or contaminated water during rainy season. During summer season they have to fetch water from a long distance

Dalit women take their bath and put on washed clothes only on the eve of festivals. It is due to consideration for ritual purity. Otherwise, they do not maintain cleanliness due to poverty. Majority of the Dalit women are below the poverty line.

Wells and handpumps in the Dalit hamlets were in very bad condition. They were not properly maintained. Women clean their utensils near to the well. The platforms are broken and surroundings are full of mud. There is no awareness of sanitation.

Muslim and Bengali Women

Most well-to-do women of Khan, Mallik and Sheikh had handpumps or wells in their houses but women below poverty line like Fakir and Ansari did not have any handpumps/wells in their houses. Some Muslim women expected wells or handpumps in the Fakir tola. People of the whole village drink water from handpumps but they use well water for washing houses and cleaning utensils and for their livestock. Purity in life is enjoined for Muslims and for this purpose they use water for *vaju*. It is noted in religious texts that whosoever is clean, will get *Sabab* (God grace).

During rainy season most of the women take boiled water for the fear of diarrhoea. Owing to contamination of water, cholera also breaks out. Muslim women always try to be neat and clean. They drink water after filtering. Some of the Muslim women take bath three to four times a week. Even frequency of washing clothes is more in Muslim houses. In the Muslim hamlet the community well or handpump was maintained well. This was not so in other hamlets. About 50 per cent government handpumps were defunct in other hamlets. Community well was maintained by the Muslim community in good condition while it is not so in other caste localities. In the Muslim hamlet the upkeep and maintenance of government wells and handpumps by the community is as good as private wells and handpumps.

(B) SANITATION AND COMMUNITY HEALTH

The views of the women on sanitation and community health in the different cultural zones of rural Bihar is very important for analysing the existing situation of sanitation and community health. The difference in the response of tribal, nontribal and minority women towards sanitation and community health is discussed below :

Tribal Women

Old Munda women hold that toilet is a source of impurity. It hurts their traditional and religious belief but young women do not regard religious factor as a constraint for having a toilet in the house. They feel that sanitary facilities improve the environment and lead to saving of time. Santhal women do not use toilets as a sanitary device. They do not like toilet in the house due to fear of the ancestral spirit (ORA BONGA). The faeces of the children are thrown in the drain or nearby field. They know the name of disease like diarrhoea, malaria, cholera but did not know their causes.

Tribal old women are not familiar with toilet while young women have already used such toilet in towns like Tamar, Khunti, Ranchi and Dumka which they visited for educational and social purposes and for employment. Munda tribal women reported that they were so poor that they could not pay for the construction of toilet. There were cultural as well as economic difficulties in the construction of latrine in the house. No Munda can construct toilet in the house until or unless the *Pahan* (religious headman) of the village permits. Munda women do not want individual toilets but prefer community toilet with the permission of the Gram Sabha.

In the tribal belt everybody accepts that scarcity of water is a crucial hindrance in the adoption sanitary measures. Old women refuse to clean the toilet while young women were ready to do so. Toilets are the felt need of young women but not of the old.

Christian tribal women who are more exposed to outside influence are ready to give land for the sanitary latrines while their unconverted sisters did not like the idea. But both the groups were not ready to pay for the construction of latrines. Unconverted women were not ready to break the tradition while for Christian women economic and technical constraints were more important. Similarly, Christian women were ready to use the manure derived from human excreta but others considered it to be polluting.

Non-Tribal Women

Upper Caste

Upper caste women were more neat and clean than the backward and Dalit caste women. High caste women take bath after four or five days while low caste women after a week during the winter season. Educated women of high and low castes take their bath regularly. Uneducated women do not know that dirty nails can create health problems. Even hair is not washed regularly. In some cases they clean it with soap or earth.

Sanitation and health condition depend upon other factors like economic status, education, caste, modernity and awareness. Poor condition of the drain and defecation in the open were common among all the castes. Only affluent families use sanitary toilets.

Pious women of upper caste did not favour construction of toilets in the houses or close to place of worship. Generally educated women were familiar with the Sulabh Sauchalaya. Upper caste women are in favour of house toilets, even though they are not affluent. They were in favour of individual toilet rather than community toilets. Non-tribal women view toilet from the point of privacy rather than that of health. Majority of the women are illiterate and do not give proper importance to sanitation. As a matter of fact, they were unaware of health, hygiene and sanitation. This is why their children defecate on both sides of the road and in the streets and drains.

Uppercaste women reported that owing to non-availability of proper place for defecation women have to face ailment like gastric pain, constipation, giddiness etc. Some times newly married women of upper caste miss their mid day meal for fear of squatting during day. When suffering from dysentery they do not take food for fear of frequent need to defecate.

Taking bath, washing clothes and cleaning the home regularly is the habit of upper caste, Christian and Bengali women. Upper caste women take their bath at the well or by bringing water in the house.

Other Backward Caste (OBC) Women

Backward caste women give precedence to the need for handpumps over toilets. Backward caste women and children are more dirty than those among higher caste of the same villages. Most backward caste women are not ready to construct latrine in the house due to poverty and small houses. Some of the backward caste like Sahni (Mallah) of Muzaffarpur district do not have their own land for housing. They are entirely dependent on the household. Khetauri (a backward caste of Santhal Paragana) women are ready to construct sanitary latrine in their houses but poverty was the main hurdle and this can be solved by the government subsidy. Some backward caste women know the name of diseases like malaria, diarrhoea, cholera but are not aware of their causes.

Some women from backward caste prefer individual latrine. Young as well as educated women realise the need for toilets and they were ready to use it while old women did not. Mostly poor women were unable to pay for latrine construction. It is not the felt need of their women, owing to illiteracy and ignorance. Young and educated women of backward castes are neat and clean.

Working women of backward caste find no time for cleaning their children. This is why the children of backward castes are more dirty than those of higher castes.

Muslim Women

Muslim women are more aware of sanitation but economic constraints come in the way of adjustment of the sanitary facilities. Poor people from Muslim community like Fakir, Julaha and others lamented that they did not have handpumps and toilets in their houses while Khan, Sheikh, Saiyad and Malik women had handpumps and wells in their houses.

Some Muslim women reported that defecation in open field is a problem of women but as head of families are male, they are not bothered about women's problems.

Most of the Muslim women were not in favour of community latrine. Muslim women always carry water at the time of going for defecation. But this was not the case with the tribal and non-tribal women. Muslim women want the construction of latrine without any subsidy while other Backward and Dalit women want it free of cost.

Dalit Women

Most Dalit women reported that they did not own the land on which they were living. They live in dirty hamlets. Dom and Musahar women rear pigs. Musahar and Dom habitations are very dirty while those of Chamar, Dusadh and Dhobi are clean. Dalit women were keen to have handpump rather than sanitary latrines in their houses. Dalit and backward women have to take bath in the river or ponds.

Dalit women do not have the land for the construction of sanitary latrine. They face social and financial problem as well as they work outside of the home for wages. They will have to return from the field for the defecation in the house. This will require more time. Mostly Dalit women felt that they can not bear the cost of the construction of the toilet.

They live in dirty hamlets and have no sense of sanitation. Faeces of children were thrown on the heap of garbage nearby houses. Children and family members wear dirty clothes. Beddings and other clothes are not clean. This is due to their poor economic condition. Majority of the children of Dom, Turiya, Musahar were very dirty. They neither take their bath nor clean their clothes regularly.

Owing to abject poverty, ignorance and illiteracy they have no sense of health and sanitation. Dalit women want community latrine as they have no land for individual latrine. But they want such community latrine to be constructed by the government.

10

ROLE OF NGOs IN RURAL WATER SUPPLY AND SANITATION

Voluntary organisations emerged as the possible alternatives because of their grassroot contact with the people, flexibility in their working strategy and commitment to serve those who are in need. These agencies have long been working in their own humble way and without adequate aid for the achievement of their objectives with their own leadership, organisation and resources. Any plan for social and economic regeneration should take into account the services rendered by these agencies and the state should give them maximum cooperation in strengthening their effort. Public cooperation through voluntary social service organisation is capable of yielding valuable results in channelizing private efforts for the promotion of social welfare.

It is necessary to involve voluntary organisation in the rural water supply and sanitation work in Bihar. The performance of Public Health and Engineering Department (PHED) in Bihar is very poor. The number of handpumps and other water facilities and the number of individual and community toilets is way behind the targets fixed by the government. It would be better to transfer the funds to the voluntary agencies to complete the task.

Sulabh International Social Service Organisation

Sulabh International Social Service Organisation earlier known as Sulabh Sauchalaya Sansthan is a pioneering NGO in the sanitation programme. It was founded by Dr. Bindeshwar Pathak as a non-profit voluntary social service organisation. It has a long experience in the construction of low cost sanitation as well as training for the construction of low cost sanitation.

In India the situation is more pathetic, deplorable and gloomy because a particular section of society is assigned duty on hereditory basis. This caste was called *Bhangi*.

In developed countries, the liberation of scavengers has already taken place and alternative methods and system of disposal of human excreta have been adopted. But, in India the problem still persists and lakhs of scavengers are still engaged in this sub-human occupation. Dr. Pathak has launched a workable scheme which has succeeded not only in liberating and rehabilitating scavengers, but also in improving environmental sanitation. The innovative low cost system of sanitation developed by the Sulabh Shauchalaya Sansthan has been adopted even by the common people with limited means. The objective of the Sulabh movement are :

* To restore human rights and dignity to scavengers.
* To prevent environmental pollution.
* To persuade people to plant trees around public toilets for the protection of environment.
* To harness non-conventional energy sources from human waste and save fuel and forests.
* To procure manure from Sulabh Sauchalaya and use it to raise farm productivity.

With the financial assistance given by the Department of Welfare, Government of Bihar under

Special Component Plan for scheduled castes through Bihar State Scheduled Caste Co-operative Development Corporation, vocational training was given to liberated scavengers and their wards.

Sulabh International constructed 1198 toilets in primary and middle schools in rural areas of Bihar. Fund was released by the Education Department, Goverment of Bihar. The year and number of toilets are mentioned below :

Toilets Constructed in Primary and Middle Schools

Year	*No.*
1986-87	489
1987-88	395
1988-89 1989-90 1990-91	314
Total	1198

The situation in rural areas is totally different from that in urban area in Bihar. Service latrines are fewer in number in the villages. This is why founder of Sulabh International Padma Vibhusan Dr. Bindeshwar Pathak did not forget the rural area and the necessity of the villages. Consequently training of village masons was conducted by the Sulabh International. But the real problem is that villages in Bihar are not having any structure like Municipality in urban area which will take the reponsibility of construction and conversion of toilets in the villages. Panchayati Raj Institutions are not effective in Bihar. If Panchayat provides the fund, toilets can be constructed in the rural areas of Bihar.

A massive programme for providing latrine in rural areas is being planned for implementation in consultation with the state government. The approach is to develop human resource and provide alternative designs of Sulabh Sauchalayas. The beneficiary is given the option to select a design which suits him most. The programme is supported by publicity, motivation, health education and follow up inputs ensuring community participation, specially women's involvement. Sulabh is implementing the programme to provide Sulabh Sauchalayas in rural areas of some districts in Bihar and Orissa.

Dr. Bindeshwar Pathak's concept for development of rural sanitation includes a five tier training strategy for creating a cadre of village based "change agents" to perform various catalytic functions and serve as a strong link between the official/non-official programme implementation agencies and the people (the intended beneficiaries).

National Rural Development Society

National Rural Development Society has constructed low cost toilets (common latrine) sponsored by the UNICEF. The districts and the blocks where the toilets have been installed by this organisation are mentioned below :

District	*Block*	*No. of Toilets*
(a) Khagaria	Khagaria	736
	Gogari	107
	Alouli	57
(b) Patna	Phulwarisharif	150
	Patna Sadar	100
	Total	1150

Khagaria Bal Seva Samiti, Khagaria

Khagaria Bal Seva Samiti has constructed 350 toilets in the Khagaria district with the help of UNICEF.

This brief survey of voluntary action in this field reveals the sorry state of things. Even though funds are available, voluntary agencies have not shown much interest in this work. Creation of awareness, advocacy and networking for this cause is essential and this can only be done by the voluntary agencies.

11

MOBILISATION FOR PEOPLE'S PARTICIPATION

The participation of users in water supply and sanitation systems, whether in rural or urban and peri-urban communities, is critical to long-term sustainability. For a water supply and sanitation system to be sustainable, users must accept their responsibilities.

In rural and peri-urban communities users normally are expected to accept a more active type of participation. They should participate as community members in the planning, construction and operation of new or improved facilities. In most such communities, on site facilities are the norm, and these lend themselves to community or household efforts. Ideally, the role of the community in both rural and peri-urban areas is to participate actively in all phases of decision-making and project development.

Experience has shown that water and sanitation projects have much greater prospects of success when communities are involved in project decision-making. In rural areas this involves planning, designing, constructing, operating and maintaining facilities. Community participation is not simply a matter of keeping the local people informed, seeking their ratification of decisions already made elsewhere, or letting them build parts of the system. True involvement means participation in decision-making and management. Anything less, will not be sustainable for a long time. Also the capacity to become effectively involved must be built over time in a community. It does not happen overnight.

Bringing about this level of community involvement requires a great deal of work—in peri urban even more than in rural areas and has substantial costs associated with it. However, when appropriate ways can be found to ensure community participation, particularly the participation of women, the benefits are great and extend beyond the project.

Community Involvement

All members of the community, not just elite and formal leaders must participate actively in the development process in order for projects to succeed. In rural areas it is found that the wealthy members of a community and those accustomed to leadership roles will participate eagerly given the opportunity, while the poor, Dalits and minorities, and those not accustomed to leadership roles (women, for example) often need special encouragement and probably special training in participation. In general, building on whatever community management structure already exists in better than attempting to create a new one. The community's informal leadership should be identified early in the development process, since it may differ significantly from the formal leadership and will be valuable in organising the whole community. Special steps should be taken to involve the very poor and other non-influential members of the community who may be ignored by extension agents because they are difficult to reach. Even these groups, however, have their own informal leaders who can be identified and drawn in to the process.

The Role of Women

In rural areas, women are the primary users of water supply and sanitation project facilities, and often provide most of the construction labour for them as well. Nevertheless, they all too frequently are inadequately involved in project development and are viewed as passive beneficiaries of improvements in infrastructure rather than active participants in project planning and utilisation. Involvement of women in project planning and implementation is essential both to sustainability and to realising long term health benefits. Women's participation in community management of rural water supply and sanitation will be more sustainable. In many rural areas, men go to the cities in search of paid employment while women are more likely to live in the community year round, making them more reliable caretakers and repair technicians.

The Role of Panchayats

The role of Panchyati Raj Institutions (PRIs) in taking up the responsibilities of Rural Water Supply and Sanitation Programme (RWSSP) to make it sustainable through proper planning and effective implementation in rural areas. Decentralization initiatives through Constitutional Amendments (73rd and 74th) to reorganise and revitalize the local governance system in the country has provided an opportunity to assign crucial development tasks to the grassroot organisations. Considering the entire process of decentralization as the necessary social prerequisites of nation-building and a means of effective democratic functioning, panchayats will be the suitable vehicle for planning, implementation, maintenance, monitoring and Information, Education and Communication (IEC) activities for this programme. The implementation of any programme is more likely to be effective and successful where the users themselves are in an identifiable group or community with its own authority structure. Local Institutions are most intended to have and to use local knowledge, to mobilize local resources, to identify local problems and ascertain the felt needs and aspirations of the people to respond quickly to change and to create a climate of opinion influencing behaviour. The local leadership will be in a position to establish and carry out systematic, participatory problem solving process. Local bodies are also responsive to the urges of the people. In this perspective, a suitable organizational framework is suggested as an alternative administrative machinery.

Gram Sabha

The basic unit of the Panchayat system is the Gram Sabha. It is the village assembly comprising all registered voters within the village panchayat area. Every State Panchayat Act has a mandatory provision to constitute this body. This forum can be utilised for the active participation of the community in Rural Water Supply and Sanitation Programme stages, right from planning to implementation management. In the meeting of the Gram Sabha, all information relating to the programme can be fully shared to make the people aware of it. This body will play a positive role in identification of problems and needs, harnessing of local resources (physical, economic and manpower) and integration of local knowledge with the technical aspects of the projects suitable, site selection for maximum social benefit etc. Similarly, awareness about environmental sanitation can be created by apprising people about various health hazards due to prevalent practices. An elaborate schedule for IEC activities can be drawn up and discussed in the Gram Sabha.

Village Panchayat

The village Panchayat will be a statutorily elected body comprising representatives of weaker sections of the society including women. Some of the State Panchayat Acts provide for Constitution of Standing Committees to deal with specific subjects. Here, it would be advisable to set-up Standing Committee exclusively to deal with the subject of RWSSO which will exercise all administrative authority for the purpose. It is further suggested to make a women representative as its chairperson. Considering the practical aspects, a woman representative will be most suited to head this committee. In rural areas, women have the most to do with water, be it collection, transportation, storage or use. In the case of sanitation also, women can be the most significantly contribute in the development of good habits of personal and public hygiene. In other words, women are the biggest suffers and can work as an effective agent of change. Women being the actual managers of domestic water and sanitation practices, management of this committee by them will be more appropriate. Also women representatives will be available due to mandatory provision for reservation of one-third of seats in the Village Panchayat. Women's representation in the committee will create the right environment for the discussion of issues of water and sanitation project.

Village leaders do have influence over people cutting across social barriers and power equation in the village. Therefore, if they can be motivated to take up water and sanitation campaign in their area, they will be able to explain the imperative need to use safe and clean water and good hygienic practices to the people.

The Village Panchayat will obtain and collect genuine household wise socio-economic and health data at the village level. All related data for the village can be generated by the Village Panchayat and can draw an action plan suitable to the area. The veracity of these data can be checked up and the picture up-dated periodically so-that one could plan measures for complete coverage. The data base will also facilitate the monitoring of the programme. Another important function of the standing committee on RWSSP will be to organise community awareness camps and exhibitions for sensitizing the people to the imperatives of safe water and clean hygienic practices. Thus the committee will be dealing with plan formulation, execution, maintenance, supervision, monitoring, awareness building etc.

For effective implementation of this programme, panchayat representative should be given intensive training so that they acquire the necessary knowledge and skill. A comprehensive training design should be prepared to impart training to all elected functionaries and other staff.

Environment Building and Social Mobilization for Sanitation

The social mobilization for sanitation project involves more than its name indicates. In fact, it concerns a communication strategy including advocacy, social mobilization and programme communication.

Advocacy consists of the organisation of information into arguments to be communicated through various interpersonal and media channels with a view to gaining acceptance from political and social leaders and preparing a society for a particular development programme.

Social mobilization is the process of bringing together all feasible and practical inter sectoral social allies to raise people's awareness of and demand for a particular development programme, to assist in the delivery of resources and services and to strengthen community participation in order to increase sustainability and self-reliance.

Programme communication is the process of identifying, segmenting and targeting specific groups/ audiences with particular strategies, messages or training programmes through various mass media and interpersonal channels, traditional and non-traditional. Advocacy consists of a large number of what are traditionally known as information and public affairs activities such as lobbying with decision-makers through personal contacts and direct mail, holding seminars, rallies and news m king events, ensuring regular news paper, magazine television and radio coverage and obtaining endorsements from popular people. The goal of advocacy is to make the programme a political or national priority that can not be swept aside with a change in administration.

Social mobilization involves the creation of a social movement for a particular programme by mobilising all kinds of allies at the national, regional and community level. It is less concerned than advocacy with direct behavioral change. Rather the aim is to create demand, in this case for sanitation improvements and thus a favorable climate for behavioral change. Social mobilisation is the glue that binds advocacy activities to more planned and behaviour oriented communication activities.

Programme communication comes closest of the three communication processes to being exemplified by what we have defined as hygiene education. Programme communication is the process of facilitating behavioral change, using all available communication channels. Successful instances of programme communication are those that are well planned and based on a sound base line and assessment of needs are specific and relevant to their target, groups and rely heavily on interpersonal communication and motivational efforts by community based workers and volunteers.

Voluntary agencies should be equally involved to mobilise people for participation involved to mobilise supply and sanitation programmes. It is very important to involve people and people's institutions in the process of conducting environment creation. Each steps of the environment creation activities needs to be planned in detail with clarity of the objectives. Similarly, a lot of flexibility should be permitted in the planning of these activities according to specific needs of different areas.

Local culture, traditions and folk elements may provide great strength to the environment creation programmes. Therefore, folk songs, dances, festival etc. can be effectively used for motivation and mobilisation of people. Participation of women, youth, cultural groups, educated and illiterate and persons from Scheduled Tribes, Scheduled Castes and minorities is crucial for creating awareness for Sanitation and personal hygiene in the community and for mass mobilisation in favour of rural water supply and sanitation programmes in Bihar. A multi media approach should be adopted where in all possible media like print media—wall writing posters, news papers etc. electronic media, Akashvani, Doordarshan, Video, Cinema Slideshow etc. and cultural forms for publicity *Kala Jatha, Prabhat Pheri, Jan Sabha*, rallies, processions etc. are used to generate maximum awareness, motivation and mobilisation for the rural water supply and sanitation programme.

12

SUMMARY AND AGENDA FOR FUTURE ACTION

The present study has been carried out in one block from each of the 13 divisions covering all the socio-cultural zones of Bihar.

The main objective of the study was to assess the existing situation of water supply, public hygiene and sanitation in rural areas of Bihar. Emphasis was also given to study the role of channels of communication, opinion leaders, mass media etc. in the spread of the programme of rural sanitation in Bihar. The role of socio-economic and cultural factors regarding sanitation and water supply is the main focus in this study.

The importance of safe drinking water, sanitation and sustainable development has been discussed. The review of studies describes the situation in the same context in Bangladesh and Pakistan. The evaluation of Past Rural Latrine programmes in India conducted by Technological Advisory Group (TAG) is also covered. Papers by Vijay Kochar on sanitation and culture have also been included in the review.

The chapter on research design delineates its the area coverage, field work schedules, sampling frame and field experience. The next chapter details the socio economic profile of the respondents.

Habits relating to water use, hygiene and sanitation in different cultural zones of rural Bihar form the contents of the next chapter. Possession of toilet, drinking water facility is also dealt with in this chapter. The existing situation of hygiene behaviour and diseases in different cultural zones in Bihar is also discussed.

The role of social variables in the existing water supply and sanitation situation in the rural area of Bihar has been brought out in the next chapter. The reason for not having toilets, attitudes towards community and women sanitary complex as well as role of channels of communication have also been discussed.

The chapters following bring out socio-cultural constraints associated with sanitary innovations, hygiene behaviour and water use pattern. Factors like male dominance in the family, purdah system, small house sites, folk practices, religion, concept of purity and pollution, among castes and tribes are also explained in this chapter.

The next two chapters reveal the views of the youth and women on water supply, sanitation and community health in rural Bihar in all the cultural zones of Bihar. The difference in the response of tribal, non-tribal and minority women towards water supply, sanitation and community health has been discussed.

The contribution of NGOs in water supply and sanitation in rural Bihar has been focussed in another chapter.

The penultimate chapter highlights the mobilization for people's participation. The role of Village Panchayat, women and environment building and social mobilisation for sanitation has been explored therein.

Main Findings

Approximately 43 per cent people do not have any drinking water facility of their own. Most of the people in the drought prone area like Palamu, Jamui and other problem areas take contaminated water from river, pond, chuan and streams. Personal handpumps were only found with only 6 per cent in tribal cultural zone, 55 per cent in Magahi zone, 79 per cent in Bhojpuri and 83 per cent in Maithili cultural zones.

Government handpumps were found with 2.75 per cent samples in tribal, 4.33 per cent in Magahi, 16.33 per cent in Bhojpuri and Maithili cultural zones. Dugwells provide drinking water in tribal zone to 87.75 per cent samples and in Magahi zone to 39 per cent to 67 per cent in Bhojpuri and to 4.33 per cent in Maithili zone. 0.67 per cent chuan (surface spring) and rivers were also used for drinking purpose in tribal zone. In Magahi zone people use river water for drinking purposes particularly in Jamui district which is drought prone area.

Majority of the sample households reported that dirty water drained out either in the street or dried at open places in the street. In Bhojpuri zone one-third of household's dirty water drained out in the street. In Maithili zone 45 per cent of the sample households used open pit for dirty water. In tribal zone, scarcely 3 per cent households disposed off the nightsoil of children in the toilet while 97 per cent people use bari field and orchard for the disposal of faeces. But the situation is different in other cultural zones like Magahi, Bhojpuri and Maithili where 12 per cent, 16 per cent and 18 per cent samples use toilets respectively. In all the zones toilet is used by the 11.6 per cent sample households for the disposal of nightsoil of children below 5 years. But majority of the households utilised open places like bari, orchard and river for such disposal. Such behaviour also depends upon the availability of water. In Jamui where there is lack of water only 2 per cent sample households dispose of nightsoil in the toilet.

It was found in the sample villages that no efforts were made for the operation and maintenance of wells, handpumps and other sources of drinking water. Kachcha well is also a problem. Villagers have to drink dirty and contaminated water from chuan (surface spring), well and river. Lack of proper supervision by the government staff was the main problem in the village.

About 85 per cent samples do not have any sanitation facility in the rural areas in Bihar. Mostly people defecate in the open field, nearby river's bank, forest and road side.

Altogether 56 per cent of the respondents felt the need of toilets in their houses. The percentage is lowest in Magahi zone i.e. 37 per cent while in the Maithili zone 80 per cent sample households voiced the need of toilets in their houses. In the tribal zone altogether 46 per cent felt the need of toilets in their house.

It was found in the rural areas that people used to defecate in the drains during rainy season. This was common for women and small children. The study reveals that disposal of garbage is done by very few people in the proper places. 1.3 per cent households in tribal zone 2.3 per cent in Magahi and 6.9 per cent in Bhojpuri and 5 per cent in Maithili zone used compost pit for the disposal of garbage. Approximately 4 per cent sample households use compost pit for the disposal of garbage. This shows that people have no idea about sanitation. Even street and open places were used for the garbage disposal. Overall 15 per cent sample households kept garbage in the agricultural field. Altogether 80 per cent sample households were least concerned about the proper disposal of garbage. Villagers reported that indifferent and non-coöperative attitude of government machinery was the major obstacle in the sanitation work.

Majority of the people reported that waterlogging and problem of drainage were the common characteristics in the Bhojpuri and Maithili culture zones of Bihar.

Women's organisations are almost non-existent in rural Bihar particularly for water supply and sanitation purposes, although organisation like Mata Committee and Jago Bahana were present in the Dumka district and Sulabh International Social Service Organisation was found in most of the districts. Research revealed that there was a demand of toilets in those areas where there was availability of water, high literacy and exposure to the urban way of life.

It has been found that it is the women who feel the need for toilets more than men. But tribal women were more concerned about the maintenance of wells and handpumps rather than individual or community latrine. There was acute shortage of water in the tribal zone which hinders sanitation and hygienic behaviour of the rural women. What to talk of bathing and washing of clothes, even drinking water was a problem.

The study highlights that no attempt has been made to involve the community in the rural water supply, sanitation and health programme in Bihar. Moreover, water supply and sanitation programme has not been associated with the other poverty alleviation or income generation programme.

Another problem in the rural area is to get trained masons to construct toilets.

Villagers reported that handpumps of MLA quota are given to henchmen of the local MLA. So poor people are unable to get such facility.

Disposal of excreta is only one component of sanitation which was included in the government programme but other component of rural sanitation like handling of drinking water, disposal of waste water, garbage disposal, home sanitation, food hygiene and community sanitation were not included in their programme.

Agenda for Future Action

Sustainable development in water supply and sanitation sector can be enhanced if an equitable gender balanced approach is taken in all phases of decision-making planning, implementation and management.

Rural Water Supply

In many parts of the Bihar like Palamu, Dumka, Jamui etc, water dries up during summer. Villagers have to drink dirty water from ponds, river, chuan, Dari and from pits dug in river bed. Most of the handpumps go out of order during summer. Operation and management of handpumps in the rural area has not been paid proper attention by Panchayats, government and NGOs. Government officials did not take interest in the repair of handpumps and wells. Moreover, there was no arrangement for disinfection of the wells during rainy season when water becomes polluted. This is why more handpumps should be provided in the rural areas and their operation and maintenance should be given to Village Committees, Panchayats or NGOs. Arrangement for cleaning the well water during rainy season is also essential.

There should be an arrangement of drainage for the disposal of waste water, otherwise waterlogging takes place and mosquitoes breed in the waterlogged areas. Consequently Malaria, Kalaazar and other diseases erupt.

The supply of water in a village is derived mostly from wells, tubewells and sometimes from rivers, canals, chuan and ponds. Adequate supply of pure water is the foremost among the important services that a community needs.

There is need for the active involvement of the people in ensuring success of the rural water supply schemes. People's invelovement would ensure transparency and accountability of the system. People should be fully involved in the schemes. They should be involved in the assessment of the situations, the definition of the problems and the setting of priorities. The first thing is to ensure people's participation is to make them fully aware of various schemes prevailing in there area.

They could be made aware through effective propagation of relevant information, increased literacy and the development of the neccessary institutional arrangements through which individuals families and communities can assume responsibility for their health and well being.

Maintenance of wells and handpumps should be transfered to people's committee constituted for this purpose. The committee should also determine the location of the new handpumps and wells so that the interest of the weaker sections of the population are kept in view.

Training should be given to select villagers for the repair of handpumps.

Appropriate guidelines and action plans for optimal and efficient operation and maintenance of installations in the rural areas need to be evolved at the earliest and strict monitoring and evaluation ensured. The PRIs can play a decisive role in this regard. But it is increasingly felt now that unless the PRIs are equipped with adequate knowhow and finance, the operation and maintenance of water supply installation may not improve. As a follow up of the 73rd Constitutional Amendment, most State government are likely to entrust the responsibility of operation and maintenance of rural water supply system to the Village Panchayats and, therefore, it would be necessary for the State Panchyat Finance Commission to explicitly indicate the norms and other financial implication and make a realistic assessment of requirement of funds and also to make suitable recommendations for provision of finance.

More handpumps should be installed in the village and these handpumps should be used in such a manner so that kitchen gardening can be started in more and more households. Thus income generating schemes will give impetus to the upkeep of handpumps in the rural areas.

Revised norms of the Eighth Five Year Plan should be strictly implemented. The revised norms are one handpump/stand post for every 150 persons. The handpumps/stand post should be installed at a distance of 250 metres horizontally and 15 metres vertical from the dwelling. All hamlets upto 100 or more population need to be covered.

Rural Sanitation

The sanitation situation in the developing countries has been characterised as shameful in world conferences. Provision of sanitation for all seems to be a long way off. The current situation which is scandalous is putting an increasing burden on the countries social and economic development plans and aggravating the already intolerable living conditions of the rural and urban poor. Statistics compiled by the World Health Organisation (WHO) and UNICEF show that nearby 3 billion people in the third world had no access to hygienic sanitation at the end of 1994 and the situation was getting worse year by year. Some ten million people a year are said to die from diseases directly related to safe water and sanitation facilities. Dennis Burner, chief of the Rural Environmental Health in the WHO has described the growing gap between those with safe water supply and those with adequate sanitation, adding that we should all be ashamed about the deplorable sanitation backlog.

Sanitation programme should be introduced rapidly where natural surrounding are favourable. North and Central Bihar is the place where open space has been reduced as compared to Chotanagpur plateau. Availability of water is also a natural factor essential for sanitation. It may be easier to take up villages which are close to towns and where literacy ratio is high. However, while selecting villages for toilets construction, care will have to be taken that there is adequate water, better infrastructure and strong non-governmental organisations to take up programme.

Maintenance work of toilets should be given to either Village Committee or Panchayats so that cleanliness and maintenance work should be done properly. Low cost technology like Sulabh Sauchalaya should be promoted in the rural areas.

Government should fix a grant and this grant should only be utilised by the Village Committees, Panchayat and only those NGOs engaged in sanitation. *Safai Mela* along with *Swasthya Mela* should be organised at the village level. Moreover, sanitation programme should be tagged with health.

Government should take responsibility for the sanitation work with the people's participation. Workshops, seminars, meetings and group discussions should be organised at the Panchayat and block level for the awareness generation and people's participation in the sanitation programme. Discussion must focus regarding health related problems in the village and strategies for the remedy. It will enable people to ponder over the environmental sanitation.

Compost pit, soak pit and Sulabh Sauchalaya should be constructed in the village and environment buliding programme for the sanitation and safe drinking water should be emphasised simultaneously.

Voluntary agencies should be equally involved to mobilise people for participation in the sanitation work. Local culture, traditions and folk elements may provide great strength to environment creation for the sanitation. Therefore, folk songs, dances, festivals etc. can be effectively used for motivation and mobilisation of people most of whom still do not understand the importance of sanitation as a part and parcel of the health.

Participation of women, youth, cultural groups, educated and illiterate and persons from Scheduled Tribes, Scheduled Castes and minorities is crucial for creating sanitation friendly environment in the community and mass mobilisation for health and sanitation.

A multi-media approach should be adopted where in all possible media like print media, wall writing, poster, newspapers etc., electronic media, Akashvani, Doordarshan, video, cinema, slide show etc. and cultural forms for publicity Kala Jatha, *Prabhat Pheri*, *Jan Sabha*, rallies, processions etc. are used to generate maximum awareness, motivation and mobilisation for rural sanitation.

The insanitary conditions associated with village life are to a large extent due to the fact that the individual does not feel a sense of responsibility towards others. The development of civic consciousness is essential for improving environmental hygiene. Such civic consciousness can be created only by general and health education. In order to secure progressive improvement in respect of personal and environmental hygiene, all people, children and adults must be so educated as to adopt and practice the hygienic mode of life and to refrain from doing what may be harmful to their own health and to the health of others.

It is very important to educate the general public about hygiene practice and pollution awareness habits. For this several tools are available to the local bodies, like organising general information campaigns regarding pollution control activities, seminars, audio-visual demonstrations handbills, advertisements, cleaning campaigns, hoardings, slides in cinema, traffic regulations,permanent notices for the information of the general public etc. Local bodies can organise information sessions by adopting any combination of these tools.

Different kinds of extension measures are needed in schools and colleges. Seminars aided with audio-visual guidance and campaigns by National Service Scheme, National Cadet Corps unit of school and college would go to a long way in creating general awareness among the student community. After this the students can even undertake educating people in the respective residential areas.

For aged rural and illiterate people particularly women and children, an entirely different kind of education campaign is needed. In such a campaign, the visuals would be more effective and repeated exposure to such visuals may result in greater retention of the message.

There is a great possibility of co-ordination with voluntary agencies for extension work. In fact, proper public relations would not only result in proper education to people. It would also make the local administration more responsive where people would develop and sustain faith in it. Women are the key to sustainable and hygienic water and environmental sanitation. They are not sufficiently motivated and utilised at present.

Women should be the first target group for a sanitation programme. The package of health education, community participation and appropriate technology should, therefore, involve women at every stage.

A sanitation education component should be included in the school curriculum and demonstration units constructed in schools, public health centres, Aganwaris, Panchayats multipurpose co-operative societies and other public institutions to ensure that habit of sanitation is cultivated. However, it will have to be ensured that these toilets are kept very clean even at a slightly higher cost.

Lack of fund is main reason for people demanding subsidy for the construction of toilets. This situation analysis shows that there is need for subsidy and loan for the sanitation programme and handpumps in the rural areas. Subsidy and loan is not available in rural area of Bihar for this. Loans should be arranged for the construction of toilets and installation of handpumps from commercial banks or other institutions. It is difficult to get community toilets constructed for the poor. So there is need of subsidy for construction of individual toilets. Community toilet is now being recognised as an essential ingredient in the adoption of a bottom-up approach where the real demand is measured by the expressed need of the end users and their willingness to pay for the cost of improved water supply. Successful water and sanitation projects must respect social and cultural norms. The project development process must involve not only engineers but also technicians, sociologists, health professionals and social workers, all interacting with the potential beneficiaries in order to create self-reliance and self-sustained action. An integrated approach is needed. Water supply can not stand alone without complementary support from sanitation, drainage, health and hygiene education.

Arrangement of training for the awareness generation regarding health and sanitation for the women is necessary because, mostly women are unaware of the diseases due to unhygienic condition of toilets as well as lack of safe drinking water.

Villagers desire that there should be a Health guide who can discuss about prevention of diseases. During the raining season, water related diseases appear but they can be checked through the health awareness.

SELECT BIBLIOGRAPHY

Articles and Books

Bendahmane Diane, B., (1993) : *Lessons Learned in Water, Sanitation and Health*, USAID, Washington.

Boot Mariek, (1995) : *Hygiene Education in Bangladesh*, UNICEF, New York, USA.

Cairncross Sandy and Feachem, Richard, G., (1983): *Environmental Health Engineering in the Tropics*, John Wiley and Sons Publications, New York.

Dumont, L. and Pocock, D., (1959) : *Contributions to Indian Sociology* Vol. III, Mouton.

Durga Prasad, P., (1956) : The Synergistic relationship between Water, Health and Development, *Kurukshetra*, Vol. XLV, October.

Feachem Richard., (1978) : *Water, Health and Development*, TRI - MED Books, London.

Khare, R.S., (1962) : Ritual Purity and Pollution in Relation to Domestic Sanitation, *Eastern Anthropologist*, 15 (2).

Kochar, Vijay, (1977) : Sanitation and Culture I : Social Aspects of Sanitation and Personal Hygiene in Rural Bengal Region, *Indian Journal of Preventive and Social Medicine*, 8.

Lall, R.B. and Seal, S.C., (1949) : *General Health Survey*, Singur, Health Centre, Calcutta.

Mishra, Sweta, (1996) : Safe Drinking Water and Primary Health Services, *Kurukshetra*, Vol. XLV, October - November.

Mohsin, N., (1988) : *Rural Water Supply : An Evaluative Study*, A.N.S. Institute of Social Studies, Patna (Mimeo).

Pathak, Bindeshwar, (1994) : *Road to Freedon*, Motilal Banarsidas Publisher Pvt. Ltd., Delhi.

Saraf, S., (1955) : The Hindu Ritual Purity and Pollution Complex, *Eastern Anthropologist,* 22 (2).

Sinha, Bakshi D. and Ghosh, Arun K., (1991): *Sanitation and Environment, Evaluation of Low Cost Sanitation in Gujarat,* Arnold Publishers, New Delhi.

Singh, S.K. (1996) : Role of Panchayats in Rural Water Supply and Sanitation Programme, *Kurukshetra,* Vol. XLV, October - November.

Tripathi, P.M., Jain, L.C. (ed.), (1985) : *'Dhamdaha' A Case Study* in *Grass without Roots,* Sage Publication, Delhi.

Verma, K.K., (1981) : *A Study of Rural Drinking Water in Bihar,* A.N. Sinha Institute of Social Studies, Patna (Mimeo).

Reports

Community Water Supply and Sanitation in South East Asia Regions. (1993) : International Drinking Water Supply and Sanitation Decade Assessment and Perspective for the 1990s, WHO, New Delhi, India.

Eighth Five Year Plan (1992-97) : Volume II, Govt. of India, Planning Commission, New Delhi.

India (1996) : Publication Division, New Delhi.

Technology Advisory Group (1995) : Evaluation of Past Rural Latrine Programme in India.

World Development Report (1993) : World Development Indicator, World Bank, Oxford University Press, New York.

INDEX